Water Consumption, Tariffs and Regulation

Water Consumption, Tariffs and Regulation

Edited by
Francesc Hernández-Sancho

PUBLISHING

Published by **IWA Publishing**
Alliance House
12 Caxton Street
London SW1H 0QS, UK
Telephone: +44 (0)20 7654 5500
Fax: +44 (0)20 7654 5555
Email: publications@iwap.co.uk
Web: www.iwapublishing.com

First published 2019

British Library Cataloguing in Publication Data
A CIP catalogue record for this book is available from the British Library

ISBN: 9781780404677 (Hardback)
ISBN: 9781780404684 (eBook)
ISBN: 9781789060515 (ePUB)

Contents

About the Editor

Dr. Francesc Hernández-Sancho has a PhD in Water Economics and is a full Professor at the University of Valencia (UV) in Spain. He is Head of the Water Economics Research Group (UV) and Director of the Master on Water Management (UV). He belongs to the excellence research micro-cluster 'Efficiency and economic and environmental feasibility for water management projects'. He is Leader of the Water Economics Working Group in the IWA SG on Statistics and Economics; he was a consultant for the United Nations Environment Programme (UNEP); is a collaborator in research activities focused in water management and water reuse projects with the Food and Agriculture Organization (FAO), and the North Atlantic Treaty Organization (NATO).

His research topics are: efficiency of water management, water economics, technical efficiency of urban water management, cost-benefit analysis in water projects, monetary valuation of water services, water tariffs, water reuse in urban areas, cost efficiency in water management, feasibility studies for water and wastewater plants and water reuse projects. He has participated in more than 40 International Research Projects related to these topics. He has also published more than 130 articles in scientific journals and more than 45 books and book chapters. He has more than 150 congress contributions in efficiency analysis and water economics. He was an Associate Editor of the journals *Water Science and Technology* and *Water Supply* (IWA Publishing), and is on the Editorial Board of *Water Economics and Policy Journal* (World Scientific).

Preface

In one of the Management Committee (MC) meetings of our Specialist Group (SG) on Statistics and Economics some time ago, the idea was launched to write a book about 'issues we are dealing with in our Group'. At that moment no one had a clear vision about what this book should contain exactly. However, Francesc Hernández-Sancho, a valued member of our MC and the leader of our Working Group of Economics, took on the challenge not knowing at that moment what stood before him or what the outcome would be. Now the book is a reality and, from our SG on Statistics and Economics, we are delighted with what has been achieved. We are sure this book will be very useful for all of us who are part of the exciting world of water.

One of the key characteristics of our SG is that we structure our program planning by organizing MC meetings twice per year in different locations around the world. Since we are in a particular location for the meeting anyway, we always ask the local host (company) if there might be interest in a workshop for the water managers in their country, completely organized and run by us. On the question of what the topic of that workshop could be, we always answer that we have a broad expertise in our Group on 'all economic, financial and statistical aspects and other aspects of business operations like efficiency, water demand, regulation, etc., in the water industry', meaning that the topic of the workshop can be chosen from a very broad panorama within the boundaries of 'economics and business operation of the water industry'. And it is exactly this that the book that lies before you now has turned out to be. The title of the book is '*Water*

Editor(s): Francesc Hernández-Sancho
doi: 10.2166/9781780404684_xi

consumption, tariffs and regulation' and in the nine chapters written by very experienced specialists from different countries you can read about the broad range of 'economic' issues facing the water industry.

The content covers issues from water abstraction, water demand, waste-water treatment and water reuse through tariffs, cost recovery and private–public partnerships to economic regulation. The authors provide you with an analysis of the respective issues, with detailed information on the subjects of their analysis, realistic solutions, or possible directions to contribute to solutions, and factual conclusions or conclusions to elaborate further.

One of the highlights you can read about is that water scarcity can have, of course, very different causes and thus very different solutions, but in any of the cases an important solution can be the use of unconventional resources like desalination or waste-water reuse. Also, a holistic approach is shown in this case advocated in which aspects of efficiency, pricing, environmental sustainability and equitability play an important role and where a sustainable, balanced and equitable water supply can only be guaranteed with proper cost recovery. Analysis of public–private partnerships all over the world shows that these PPP's depend very much on local/regional circumstances. The same local and regional differences apply to the different types of economic regulation and this is – if in use – is always considered as a substitute for lack of competition and can prove to be a strong instrument to avoid the risk of inefficiency.

You may conclude from the above that this book offers valuable insight into different kind of issues written by recognized experts in the water industry. This can help water managers all over the world, who face these kinds of problems daily, to better understand what they are dealing with and to better substantiate their decisions.

And so that is why I would like to thank, on behalf of the members of the Management Committee of our Specialist Group on Statistics and Economics, all the authors of the different chapters for providing their great expertise and for sacrificing their limited free time.

I especially would like to thank my good friend Francesc Hernández-Sancho for all the time that he spent in the creation of this book and for his perseverance in finalising it. I know that it has cost him the necessary drops of sweat and extensive efforts to find the authors, to coordinate their activities and to do the final editing. It is very much appreciated!

Ed Smeets
Chairman of the Specialist Group on Statistics and Economics of the International Water Association

Chapter 1

Overview of water abstraction in the world

Águeda Bellver-Domingo and Francesc Hernández-Sancho

Water Economics Group, Department of Applied Economics II, University of Valencia, Avda. Tarongers s/n, 46022 Valencia, Spain

Corresponding author Email: Agueda.Bellver@uv.es

ABSTRACT

Water is an essential resource to ensure the continuity of both population and current productive systems. A lack of management in water abstraction is partly responsible for the decrease in water availability, but climatic influence on the hydrological regime must also be taken into consideration, exacerbating water stress conditions in arid and semi-arid areas. Countries need to face this alarming situation in which the universal right of access to drinking water is jeopardized. Thus, lack of water is no longer just a problem in developing countries, but affects the entire world population. There are several ways to solve the problem. This paper analyses water abstraction across the world, drawing on the different response options to water scarcity, relating to the new management of the water cycle in which non-conventional water sources become the efficient response to solving the water stress problem.

1.1 INTRODUCTION

Efficient water use is a key aspect given the influence water management and climate change have on climatic dynamics in relation to the balance of the

Editor(s): Francesc Hernández-Sancho
doi: 10.2166/9781780404684_0001

hydrological system. This influence results in overexploitation of water sources and exhaustion problems. Current trends in the planning and management of water sources are changing, because it is becoming evident that excessive consumption of water resources directly affects water resource availability, both on the surface and groundwater. This damage is transferred to ecosystems and the living beings that inhabit them, causing serious impacts on biodiversity. If water sources are not protected, pollution and overexploitation would reduce the amount of water available for urban water supply and ecosystem maintenance. Decision makers need to know how much water is available (in a basin) as a starting point for managing a water supply and consider an integrated and multidisciplinary approach, in which not only anthropogenic uses are considered, but the maintenance and protection of the ecosystems which also depend on that water resource (World Bank, 2010).

Over time the demand for water will continue an upward trend because the world population will increase. The first measure for any government is to increase water supply, build more dams to retain water or look for water sources from unconventional origins to meet demand. Although this book focuses on the urban aspect of water, bear in mind that population increase comes with an associated growth in food demand, which causes increased water consumption in the primary sector. Thus, water management not only requires a purely urban focus but a multidisciplinary approach, where different water uses and demands are combined with existing and potential sources of water (European Environmental Agency, 2015a). The aim of this chapter is to show the main water sources for water supply and their influence in the different regions of the world. The first section shows the different water sources according to their origin, the second section includes different examples of water abstraction by country, and the third section shows the evolution of water withdrawals by country and by source.

1.2 OVERVIEW OF WATER ABSTRACTION ACCORDING TO SOURCE

How much water is there in the world? How much water is being consumed? The answers to these questions are well known. Total water available for humans is only 0.3% of lakes and rivers and some of the water stored in aquifers (GWP, 2012). This 0.3% is water that is part of the hydrological cycle: runoff of rainwater and evaporation and evapotranspiration processes. The average annual rainfall is 119,000 km^3, of which 74,000 km^3 is evaporated and 45,000 km^3 flows into lakes, rivers and infiltrates soil. However, actual water volume available shows that, of those 45,000 km^3 that flow through the earth's surface, between 9,000 and 14,000 km^3 correspond to water volume that is economically usable by humans. However, 3,600 km^3 is accessible to humans to meet demand, i.e., this volume can be extracted annually by available technologies. These numbers show that there is a low amount of water available for human consumption. Furthermore, the amount of water stored is at risk due to different environmental impacts and not only human consumption has to be considered but also the ecological flow of water ecosystems. The water situation is delicate, regardless of whether it refers to the present or the future. Water availability is highly influenced by the rainwater distribution, both under natural weather conditions and global warming conditions. This situation causes imbalances between different regions (water stress or extreme flood conditions). Furthermore, increasing water demand, human intervention in water ecosystem and geomorphology exacerbate the problem (FAO, 2002). Bearing in mind that extreme weather events may cause severe droughts and floods that affect water availability, authorities need to efficiently manage their water resources. Regional water scarcity is difficult to address because the demand for water is always increasing. In order to efficiently manage the water volume and reduce its vulnerability, each region needs to know the water volume available and the climatic and geomorphological conditions that affects water sources, such as identifying rates of renewal and recharge, the types of the aquifer and so on (GWSP, 2015).

There are two main sources for human water supply: surface and groundwater. The proportion in which they are used varies according to the region and the availability of resources and weather conditions. Despite this, the main source used is surface water, followed by groundwater (Richey *et al.*, 2015). Water sources need to be protected, correctly managed, and regulated in order to avoid overexploitation (WHO/UN-WATER, 2014). Overexploitation and pollution are the two most serious problems that surface water faces. In many places there is only low surface water volume in riverbeds, causing serious problems to the surrounding ecosystem which lacks the water volume necessary to support all biological processes (Palaniappan & Gleick, 2009).

Groundwater is one of the most in-demand resources in places where surface water availability is reduced. In addition, in areas with strong water stress problems it is an important source, despite the obvious risks of overexploitation and marine intrusion that can occur in coastal aquifers (Gössling, 2001; Richey *et al.*, 2015). It is usually thought that groundwater is the water source less used, but in countries like Germany, Netherlands, Austria, Finland and Sweden the opposite is true, the water used for urban supply is groundwater because of its high availability and ease of withdrawal. In overall data, groundwater is the drinking water source for two billion people (Richey *et al.*, 2015). The advantage of using groundwater is its higher quality and its low cost of treatment in comparison with surface water, as happens in all countries of southern Europe (EUROSTAT, 2001). In contrast, the main disadvantage of groundwater is extraction cost, moreover the costs associated with monitoring because an aquifer is a geologically complex system influenced by numerous variables that determine the recharge water flow and overall water levels. This structural complexity makes water stress problems on aquifers more difficult to solve and if the water volume stored is reduced, this can result in serious socio-economic and environmental consequences (Richey *et al.*, 2015).

Worldwide groundwater volume for drinking water supply is around 10.5 million km^3, however the ability to access these sources is often limited (investment, operational and maintenance costs). Approximately 35% of the subsurface layer of the Earth is aquifers, because layers of sedimentary geologic formations are optimal for storing water. This is the type of surface aquifer (shallow aquifer) has afforded the supply of drinking water for a long time. However, there are also aquifers that exist in complex hydrogeological structures (18%) where water withdrawal work is absent or requires major financial investment for drawing water (WHYMAP, 2008). Other problem of complex hydrogeological structures is that there are some aquifers that are fossil (non-renewable groundwater), i.e., a reservoir of water that formed in the past whose supply is interrupted, usually by a geologically extreme phenomenon, isolating the system. Water has been preserved over time and is perfectly suitable for drinking purposes, but when the aquifer is overexploited recharge is not possible (Gössling *et al.*, 2012; Palaniappan & Gleick, 2009).

Table 1.1 shows groundwater area data by continent. In all continents, the largest area is covered by shallow aquifers, which explains why groundwater is the second most used source of water. From Table 1.1, in a global context, the largest groundwater basins occupy 47.32 million km^2, of which 47% are shallow aquifers while only 18% are aquifers in complex geological structures.

Groundwater withdrawals will increase in the future. Water stress conditions, population and productivity growth and changes in weather patterns are the main vectors for the increase in water consumption. For these reasons current water

Table 1.1 Aquifer distribution data by continents and world.

	Major Groundwater Basins		**Local and Shallow Aquifers**		**Complex Hydrogeological Structures**	
	Million km^2	**%**	**Million km^2**	**%**	**Million km^2**	**%**
Africa	13.48	44.9	13.22	44.1	3.31	11.0
Asia	14.54	32.0	22.98	50.7	7.84	17.3
Australia, New Zealand	2.60	32.5	2.49	31.1	2.90	36.3
Europe	5.15	53.0	2.74	28.2	1.82	18.8
Central/South America	8.35	45.0	8.18	44.1	2.02	10.9
North America	3.21	15.0	12.40	58.1	5.75	26.9
World	**47.32**	**35.6**	**62.02**	**46.6**	**23.64**	**17.8**

Source: WHYMAP (2008).

volumes have to be monitored while supply and demand for water are planned (Richey *et al.*, 2015).

Planning of water supply to meet demand has been happening since ancient times with the construction of dams and channels, for example, the Romans and their aqueducts. The result is that humans have been modifying the hydrological cycle to store more water, thus ensuring their survival regardless of geographical area. Construction of dams, channels, and other regulation infrastructures has been a constant solution for the different water uses and not only has the number of these infrastructures increased, but also their storage capacity (FAO, 2000). The closest case that exemplifies this behaviour is the colonization of the western United States where water infrastructures modified the hydrological cycle to increase the stored water to meet both the demand and the irrigation capacity. The biggest dams were built between the years 1930 and 1965 with the aim of promoting economic, population and industrial development. However, the socio-economic change was already happening years before with the mechanization of agriculture and the development of farms: between 1850 and 1900 United States went from 1.4 million to 5.7 million farms. Colonization of the west of the country was marked by aridity problems, hence, water was the fundamental pillar for economic and social success. As a result, engineering works were significantly developed to allow snowmelt water and rainwater to be stored for agricultural, human and industrial uses downstream. By the late nineteenth century many small streams were diverted for agricultural use, but the amount of water from them was not enough to support the entire agricultural industry that was being created. The arrival of President Roosevelt changed the concept the United States had of the river, looking for its usefulness to humans; i.e., fresh water would be a commodity whose regulation and channelling were essential. This explains why the projects developed from that time were large-scale dams for irrigation and drinking water supply of the entire population, "to make the desert bloom". The era of big dams began in the 1930s seeking to generate hydroelectric power and ensure the growing demand for water. An example of these new water aims is the Hoover Dam (Colorado River), which was the first major dam built for agricultural use, which allowed population growth of Los Angeles city. Why was the Colorado River the first river with a large dam? Its riverbed was irregular allowing for hydroelectric power, mainly during flooding season where water flow increased significantly. Once the Hoover Dam was constructed, the advantages of such infrastructure in all areas of society were obvious, hence, the artificial river regulation became a priority to ensure that these achievements could be extended over time (U.S. Department of the Interior, 2005).

Building dams with greater capacity and high rates of extraction of groundwater is not the answer to meet current and future water demand. The current water situation causes insecurity of water supply and the available water varies each year as does the amount of water demanded by the cities (Scarborough *et al.*,

2015). Water regulation infrastructures are necessary to ensure continuous and sustainable economic and social growth over time. While it is true that the environmental impact associated is high, between 2001 to 2008 United States built 81 dams for water supply, and it cannot build more as a result of the severity of federal environmental regulation (Chowdhury *et al.*, 2013). An in-depth analysis shows that the United States is one of the countries that consumes the most water in the world. Considering the period 1950–2005, population density has reached 300 million people (Figure 1.1), thus the water volume supplied to households must increase. In 1950 the volume provided to the drinking water network was 53 km^3/day while in 2005 it was 167 km^3/day (USGS, 2014). Data supporting this increase in water withdrawals to supply within the period considered is shown in Figure 1.2, and this increase is due to construction of new reservoirs and channels. In general, the USGS quantifies total freshwater withdrawals as 1,321,862 hm^3/day, of which, 167,315 hm^3/day (13%) are for urban supply. Specifically, it comes from a combination of surface water (112,048 hm^3/day) and groundwater (55,267 hm^3/day).

Water abstraction has guaranteed the survival of human populations over the centuries, although previous abstraction rates allowed the recharge of sources. Currently extraction rates are accelerating, being greater than the capacity of the ecosystem to recharge the volume used. Overexploitation is a problem with dual implications: (i) environmental impacts which require a guaranteed sustainability of water in the long term; (ii) social problems resulting from ensuring universal access to water. The above example of the Colorado River, shows the evolution

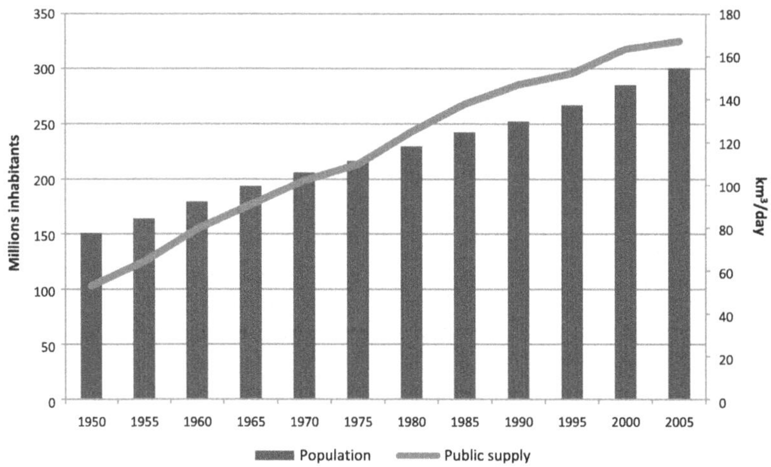

Figure 1.1 Growth of United States inhabitants from 1950 to 2005 according to water-supply volume growth for urban public supply network. Source: Own elaboration from Kenny *et al.* (2009) data.

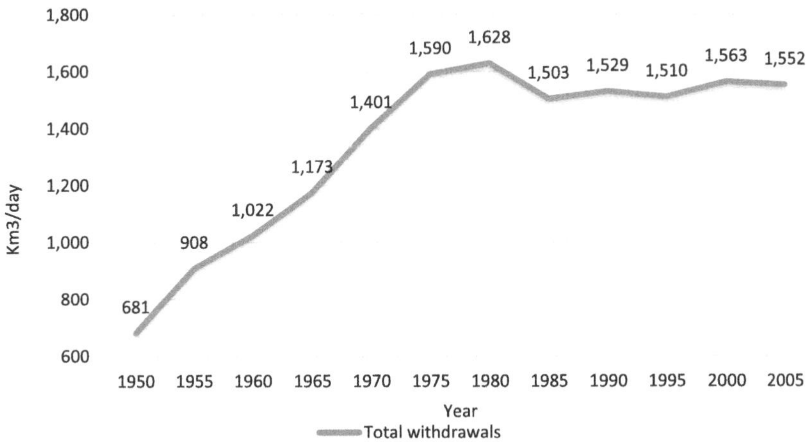

Figure 1.2 United States total withdrawals from 1950 to 2005. Source: Own elaboration from USGS (2014) data.

that has followed all societies, in which they seek to increase artificially the volume of water available to ensure downstream water supply; in fact, the current source for water supply in United States is surface water (Chowdhury *et al.*, 2013). It is the regularity of a water supply which provides confidence to inhabitants and decision makers to progressively increase water demand. This increase provokes the search for new water sources and new places for storage. An imbalance between supply and demand is created, related to the limited speed of recharge of the water system, which brings a significant economic cost associated with each additional unit of water that will be stored. The solution to this problem is to further modify the water cycle, either without considering the natural water flow (current widespread option) or to consider recharge times and natural dynamics of water ecosystems (future option) (FAO, 2000). Efficient use of water is important in achieving integral water-cycle management, especially in agriculture where savings could give more value to agricultural products with low water consumption, an area which currently represents 70% of total water withdrawals (Gössling *et al.*, 2012). Saving water in agriculture can avoid environmental problems related to pollution or high withdrawal rates of freshwater (FAO, 2002). Given that water scarcity has a regional character, for example, population growth, tourism, industrial and agricultural development along with the scarcity of soil and the long-term problem of dams as they are filled (Scarborough *et al.*, 2015), non-conventional water resources are beginning to gain importance. In this chapter, two of these resources will be discussed: reclaimed water and desalinated water.

Treated water discharged from effluent of wastewater treatment plants (WWTP) reaches the ecosystem and returns to the hydrological cycle. However, this water is not always considered within the planning of uses, processes and demands of decision makers. The volume of treated water is high, but until relatively recently it was thought that it could not be useful. The current trend is to consider WWTP as a potential source of water for environmental and agricultural uses, transforming the effluent into a new source of water (Hernández-Sancho & Sala-Garrido, 2009; Simons *et al.*, 2015). As a result, quality requirements for the effluent are positive both for agricultural or industrial use and to mitigate environmental impact (Molinos-Senante *et al.*, 2011). The reclaimed water can be used in three different ways: (i) direct use, where the effluent is used just as it leaves the WWTP, without post-treatment, for another use or for a use similar to the original; (ii) planned indirect use, where the treated water returns to the environment to maintain ecosystem flows; and (iii) unplanned indirect use, where water is returned to surface water bodies, which are a source of water for human use downstream (Wiener *et al.*, 2016). Therefore, the use of reclaimed water is a very positive way to reduce the volume of water consumed by agriculture while maintaining the environmental conditions in rivers, lakes and wetlands. This enables society to advance to the integrated management of the water cycle, whose dynamic is heavily anthropocentric. Hence, all the elements involved need to be considered. While it is true that reclaimed water is a significant step towards a global mind-set change regarding water conservation, not all countries approve that reclaimed water could be used for human consumption. In this chapter, reclaimed water has been included so that the reader may be aware of current trends in water cycle management. Reclaimed water can become a solution to the environmental problems associated with water scarcity, in that it can be used to maintain water flow requirements of ecosystems, reducing the impact of high water demand. China is an example of the use of reclaimed water in its cities, and in the case of Tianjin 80% of its water demand is met using reclaimed water, because Tianjin's government wants achieve an integrated reclaimed water-use plan for proper management of their resources and develop different methods to finance infrastructure (Gu *et al.*, 2015).

Desalination is already an answer to the problems of water scarcity in coastal areas with high population density and is of importance to the tourism sector. Water sources that can be used in the process comprise 60% of seawater, 21% of brackish groundwater and the rest comes from surface water (Ghaffour *et al.*, 2013). Desalination is an option to increase security in water supply, which has the advantage of the high quality of the obtained water. The main disadvantage has been the high cost of water production due to the high operational costs (Scarborough *et al.*, 2015). Desalination began in California in 1930 due to the severe drought that they were suffering; however, the construction of the first large desalination unit had to wait to 1950 in Kuwait which it had a capacity of 450 m^3/day. Drought conditions suffered in United States together with increases in urban population and industrial activities were the reasons that led the government to approve programs to finance the development of desalination technology (Haddad, 2013). From then, the technology used for desalination has been changing as research advanced, always looking for the best technology to improve water production while reducing operating costs (Delyannis & Belessiotis, 2010).

Production costs of water desalination have always been a barrier to the spread of this technology. However, membrane technology was not viable on the market until the early 1970s, when reduction in production costs became clearer, because before this time seawater desalination cost about US\$2.10/m^3. As the market for desalination grew and more companies and organizations were interested in the technology, water desalination achieved a qualitative leap in reducing production costs. Overall, the price of seawater desalinated dropped to US\$0.50/m^3 for reverse osmosis plants with large capacity (large scale plants) and to US\$1.00/m^3 for desalination technology based on the distillation stages (multi-stage flash distillation). These values reflect rapid reduction in production and functioning costs. This does not mean that water desalination is a cheap option to solve the problems of water scarcity, mainly when the following process inputs need to be taken into account: (i) power consumption; (ii) desalination process configuration; (iii) plant size; (iv) geographical location; (v) raw water quality and final water quality required (post-treatment); (vi) reject discharge type; and (vii) water storage. Taking the example of power consumption for reverse osmosis, the minimum energy required is 0.706 kWh/m^3, but in practice the energy range is between 3–4 kWh/m^3 and can reach 9 kWh/m^3; although this may include energy recovery stages that reduce overall energy consumption (Haddad, 2013). Despite reduction in energy consumption mentioned above, power consumption forms 30–50% of operating costs, quantitatively speaking energy is the main cost of desalination (Burn *et al.*, 2015).

Currently, process improvements are allowing desalination to become a viable option for supplying water to both coastal areas with water stress, and areas where aquifers are polluted (Ghaffour *et al.*, 2013). As current limitations of soil in coastal areas (highly influenced by urban development and tourism) prevents the construction of large water-storage infrastructure such as dams, desalination is an efficient and effective long-term response to solve water scarcity (Scarborough *et al.*, 2015). Countries like Qatar and Kuwait have already achieved 100% of the drinking water supply using desalinated seawater, and in Australia and Spain desalination is being used for irrigation and domestic uses (Ghaffour *et al.*, 2013).

Conventional water sources are no longer able to support current population growth and the attendant increase in water demand, whether for urban, agricultural or industrial uses. This means the efforts to store more water are increasingly expensive, complicated and harmful for the environment. We have the possibility of use alternative water sources (desalination and reuse), whose viability is proved but whose spread has been limited. However, the severity of droughts in arid and semiarid regions is changing the social trend towards sustainable water consumption, making humans aware of the need to use all water sources available and the necessity to improve the efficiency of water use, without forgetting the importance of government planning for the uses and demands for an entire water watershed.

1.3 ANALYSIS OF WATER ABSTRACTION BY COUNTRY

This section will discuss how water withdrawals have developed in different countries and seek to draw a global trend including both developed and developing countries in different climatic regions of the world. This will help the reader to know what actions each country has taken to meet water demand and the future trends of water supply solutions.

1.3.1 United Arab Emirates

Water sources in the area of the Middle East are very susceptible to hyper-arid climate, resulting in no viable use of surface water and aquifers having very low rates of recharge water that limit water abstraction. However, since the early 1980s, agriculture has undergone a major expansion thanks to the extraction of groundwater for irrigation, with the disadvantage that the volume of stored water has reduced and caused an accumulation of salts in the water, and subsequently a loss of quality. In general, groundwater is the source most used in agriculture: 64% of all water used comes from the aquifer; while the 30% of the water used is desalinated water and the remaining 6% is reclaimed water and is used for urban water supply (Environmental Agency Abu Dhabi, 2014). These non-conventional water resources have become a fundamental pillar in the hydrological cycle of the region, because suffering water stress forced increase efficiency in the seawater desalination process and improved WWTP with modern tertiary treatment systems that allow using water for various uses (McDonnell, 2014).

Abu Dhabi has one of the highest per capita water consumption rates, i.e., in 2008, water consumption oscillated between 565 and 920 l/capita/day while the world average ranged between 160 and 220 l/inhabitant/day. Aquifer overexploitation provokes salinity quality problems: in 2011, 80% of the groundwater was saline, 18% brackish and only 3% was freshwater. In 2008, the water scarcity index in the Emirate of Abu Dhabi was 33 m^3/inhabitant/year. Therefore, renewable fresh water resources in the country are far below the water scarcity threshold of 1,000 m^3/inhabitant/year (Environmental Agency Abu Dhabi, 2014). Despite these data, the inclusion of non-conventional water resources has allowed Abu Dhabi to become a green landscape in the desert challenging the natural climatic conditions of the region (McDonnell, 2014).

1.3.2 Sub-Saharan Africa

This region is undergoing strong population growth with the problem that urban settlements do not meet the necessary conditions to ensure proper water supply because many of them are informal settlements. In 2012, less than 70% of the population had access to drinking water because bringing water to homes is a very complicated task requiring large investments in infrastructures (Marson & Savin, 2015). Drinking-water allocation is not centralised, and many inhabitants have no acccss to a drinking-water supply network, hence, they are obliged to look for other forms of self supply such as wells, surface water, vendors, and illegal connections. Of all people without access to water, 85% live in rural areas, so depending on a source of surface water is difficult because settlements are widely dispersed (MacDonald & Calow, 2009). For that reason, groundwater is the most used in households because it is affordable and there is availability. The problem with this supply model is that it has not adapted to the population growths because it has not been subject to urban planning, thereby causing long-term problems, and groundwater areas vary across the continent owing to the geological characteristics (MacDonald & Calow, 2009). Furthermore, groundwater is also highly susceptible to faecal pollution and, as a result of there being no suitable planning of the water supply, there are no sanitation measures or water treatment planning, hence, the faecal contaminants resulting from improper management of urban waste causes health problems from drinking polluted water (Okotto *et al.*, 2015). Despite all the risks that face groundwater, aquifers are the most feasible and efficient way to provide water to the sub-Saharan region (MacDonald & Calow, 2009).

1.3.3 China

China has undergone rapid socio-economic development that has been strongly linked to the current water shortage. The size and geographical location of the country means rainfall is not uniform in either time or space. Therefore, the infrastructures cannot store all the water needed for the population and for agriculture (Jiang, 2015). Surface water is the majority source for water supply in China, taking into account the high rate of precipitation. According to the Ministry of Water Resources reports for 2007, the total amount of annual precipitation is 5,741 billion of m^3, but total amount of surface water resources in China was 2,376.4 billion of m^3. Volume of surface water availability in northern part of China was 407.9 billion of m^3 and in the southern part of China was 1,968.6 billion of m^3. Hence, available water resources per capita were 1,869 m^3 in average and water abstractions was 578.9 billion of m^3 for urban water supply, of which surface water was 470.4 billion of m^3, groundwater 105.8 billion of m^3 and other water sources 2.7 billion of m^3, or in other words, surface water accounted for 81.2% of the total water supply, groundwater 18.3% and other water sources 0.5%. In order to increase the availability of water for the population, the Chinese government has chosen to increase the supply of water (whether for urban, agricultural or industrial uses) through the creation of new water storage infrastructures (MWR, 2012).

1.3.4 India

Rainfall in India has a marked seasonality affecting the dynamics of their water bodies, both surface and groundwater. The average annual rainfall in the country is 117 cm/year (including snowfall). The major volume of rainfall water is concentrated during the monsoon months, during which precipitation ranges from 750 to 2000 mm/year depending on the area. Climatic influence not only affects the rainfall volume registered, but the evaporation percentage is significant due to high temperatures, hence, the total amount of water available per year (after deducting for evaporation losses) is 1,869 billion of m^3, of which only 1,123 billion of m^3 is usable (690 billion of m^3 come from surface water and 433 billion of m^3 come from aquifer). The significance in the water availability (both surface and underground) is relevant from the point of view of the high population growth that the country is experiencing, which associates with an increased water demand, the supply of which will depend on increasing the volume of water collected in terms of the implementation of structural measures (dams, for example) (MWRI, 2011).

According data of Indian Government, water availability per capita for 1951 was 5,177 m^3, whereas for 2001 was 1,820 m^3. In only in 50 years water availability has fallen by 35% while population suffered a strong increase, by almost 700 million inhabitants. This trend is worrying because the projections for 2050 show that water availability per capita will reduce to 1,140 m^3, i.e., there will be a significant increase in population and associated water demand, but with the problem that the amount of water available will be less, resulting in an imbalance of water supply and deficiencies in food production. Are water resources suitably managed in the country? It is a complicated question, on the one hand the current growth of Indian agriculture needs a high volume of water (83% of all freshwater is used for irrigation) and, on the other hand, around 15% of groundwater bodies in the country are in a state of overexploitation. The solution is to take advantage of the high volume of rainfall water in the monsoon months (this period generates 80% of the total surface water runoff of the country) using strategies to store this water: in 2011 the storage capacity created was 225 billion m^3, with a projection to create 172 billion m^3 of additional capacity in the near future (MWRI, 2011).

1.3.5 Australia

Australia is suffering a serious water crisis characterised by the climate and strong water demand. Around 90% of the Australian population lives in urban areas, and 61% are located in larger cities. These data, along with a strong drought situation, complicate drinking water supply (Horne, 2015). In the year 2000 the Australian government began to change water management policies due to the severity of drought (Radcliffe, 2015).

The influence of El Niño on the Pacific coastal region causes severe cycles of drought that seriously affect agriculture and urban water supply. The last drought lasted from 1997 to 2010 (Ejaz Qureshi *et al.*, 2013). The Australian climate is responsible for precipitation concentrated in the north of the country, with lower population density and little agricultural land. The southern region is the most densely populated and has strong agricultural activity, but it is the area most affected by rainfall shortages and by the high rates of water withdrawal. The fastest government response is to establish water-use restrictions (Australian Government, 2015). The water restriction policies are a short-term option to preserve the low water volume that is available (Chanan *et al.*, 2009). The government's response has been, in most cases, to establish restrictions on the use of water in gardens to 50%, in order to avoid significant reduction in levels of aquifers and dams (Hurlimann, 2009). Historically these cities have been supplied by dams during drought periods because the population was low and supplementing with the use of surface water groundwater has also been a widely used. As has happened across the entire world, Australian groundwater abstraction has increased significantly, from 1983 to 1996 the use of aquifers rose 90% (Walsh *et al.*, 2014). Under this situation, current water volume is insufficient to meet demand and desalination is seen as a good option to increase the supply of water in the (near) future.

The introduction of desalination as non-conventional source of water would help to maintain water supply during drought times and would soften the water needs of strong population growth (Scarborough *et al.*, 2015). Due to the serious situation, desalination is not the only source of non-conventional water, the central government is also promoting the use of reclaimed water for urban and agricultural use. However, dissemination of reclaimed water is still limited, today only the 4% of overall water used is reclaimed water (Hurlimann, 2009). This new approach to water management has been successful in Australia, and in cities like Queensland, and has significantly reduced dependence on conventional water sources for urban supply, increasing the volume of stored water and ensuring long-term continuity of resource (Gurung *et al.*, 2015).

The Murray-Darling Basin is an example of water stress. It occupies a million square kilometres, is home to 16 wetlands included in the *Ramsar Convention* and supports two million people (Docker & Robinson, 2014). Water volume is low due to low rainfall, high evaporation rates and urban demand. Currently the total percentage of water stored in the basin is 38.1%, while for the year 2015 it was 47.2% (Australian Government, 2016b). This water stress situation is not new and since 2007 the Australian government has been implementing measures to mitigate the imbalance of basin water flows (Docker & Robinson, 2014). This situation involves increased salinity in the basin (two million tonnes of salt flows down the Murray-Darling basin). Strong economic investment is needed to remove salts and not damage the ecosystem (for 2003 it was AUS$305 million; Australian Government, 2013).

Searching for solutions to alleviate the drought situation has been the leitmotif of the Australian government. Desalination has become the most widely used alternative to alleviate the situation (Scarborough *et al.*, 2015). For the year 2011–2012, the volume of desalinated water exceeded the volume of rainfall in coastal areas (Australian Government, 2014). Another option is to reuse treated water for reducing dependence on surface water. Australian government data for desalinated water show the scope of these non-conventional water sources (see Table 1.2). The volume of desalinated water generated depends on which part of the continent the desalination plant is located, considering that its use is for the urban water supply. Treated water reuse, however, is more homogeneous throughout the continent, shared between environmental and irrigation purposes and urban water supply.

Reducing the scale of our analysis, we will show the case of the city of Perth (located in the state of Western Australia). Perth has been selected because surface water levels are low, so the Australian government completes the urban water supply using non-conventional resources (Table 1.2). Water bodies that supply the city of Perth have a lower water level. Table 1.3 shows the volume of those bodies of water used for the supply of Perth between the dates from 24th to 25th February 2016 (Australian Government, 2016b).

The situation is alarming and, regardless of the scale used for problem analysis, the Australian water situation is serious. The inclusion of non-conventional resources as a source of water alleviates the pressure exerted on water resources. However, there is still a long way to go to avoid water stress, especially when droughts are recurrent in time and long-term length (Radcliffe, 2015; Docker & Robinson, 2014). Australia is presented as one of the best examples of government response to water stress conditions, ensuring drinking water supply

Table 1.2 Desalinated and reclaimed water volume for Australian states.

	Desalination		**Reclaimed Water**	
	Annual Production (ML/year)	**Main Use**	**Annual Production (ML/year)**	**Main Use**
Northern Territory	475		600	
Queensland	109,311	Urban	226,711	Environ. & irrigation
New South Wales	144,215		281,440	Urban
Australian Capital Territory			32,420	Environ. & irrigation
Victoria	214,340		112,337	Urban
Tasmania			21,413	Urban
South Australia	179,693	Urban	107,285	Environ. & irrigation
Western Australia	232,230	Urban	156,846	Urban

Source: Australian Government (2016a).

Table 1.3 Stored water volume of Perth water bodies (Australian Government, 2016b).

Water storage	Capacity (ML)	Current Volume (ML)	% Full
Canning	90,289	20,815	23.1
Churchman Brook	2,204	427	19.4
Mundaring	46,896	3,544	7.6
North Dandalup	73,830	28,259	38.3
Samson Brook	7,902	651	8.2
Serpentine	136,658	29,571	21.6
Serpentine Pipehead	2,166	1,645	76.0
South Dandalup	125,073	1,053	0.8
Stirling	54,289	6,362	11.7
Victoria	9,417	1,972	20.9
Wungong	56,365	19,896	35.3

Source: Australian Government (2016b).

for its population and ensuring water supply for the suitable functioning of the productive sectors. To achieve better management and water savings, it is necessary to implement limits to water consumption. The Australian Government has called this Sustainable Diversion Limits, which allows regulation of anthropogenic surface water uses so that more water volume will be available for human and environmental purposes. This is achieved by improving the efficiency of supply systems (reducing water losses in the urban network and irrigation systems), whose water savings provoke the largest volume of water available to maintain environmental flows and reduce the impacts associated with the high demand for water (Australian Government, 2014). In addition, the Australian Government has implemented water restrictions in metropolitan areas with the aim of increasing surface water storages, mainly in the South West Coast region. This is important as groundwater withdrawals will increase in the near future as a result of drought cycles (Australian Government, 2015). Another measure for increasing water supply is desalination water, the production volume of which was larger than coastal rainfall in the first half of 2011–2012 (Australian Government, 2014).

1.3.6 Mediterranean region

Mediterranean region has several and serious problems: water scarcity, concentration of economic activities in coastal areas, and reliance on climate-sensitive agriculture. It is further characterized by being one of the regions most affected by anthropogenic alterations in rivers, such as dams

(García-Ruiz *et al.*, 2011). This is because the climate is strongly seasonal, with dry, very warm summers and cool, wet winters which causes rains only in spring and autumn, but these rains are torrential (Collet *et al.*, 2015). Furthermore, since 1970, an increase in air temperature of almost 2°C has been recorded in south-western Europe, as for rainfall, precipitation has increased in the northern Alps, yet decreased in southern Europe, where a 20% drop in rainfall has been recorded. In other words, the Mediterranean region is very vulnerable to climate change which is responsible (along with population and tourism development) for the current conditions of water stress (Collet *et al.*, 2015).

Water demand is high in the Mediterranean region where average annual demand for freshwater shows that southern countries of Europe are the most water-stressed, 180 million of the Mediterranean population has below 1,000 m^3 per inhabitant per year, and the forecast shows a future increase in those data to over 250 million within 20 years. Why does this occur? Mainly because this region is undergoing intensive demographic, social, cultural, economic and environmental changes combined with the growth of coastal settlements and tourism which generate significant environmental pressures, such as air and water pollution, waste generation, land consumption, and degradation of habitats, landscapes, and coastlines. Since 1995, tourism has grown by almost 75% and projections show that the number of arrivals could reach around 640 million by 2025 (European Environmental Agency, 2015b).

Renewable water resources within the Mediterranean region represent 1.2% of the world's renewable water resources, i.e., around 550 km^3/year. However, they are unevenly distributed, in north of region the precipitation is greater than in the south, hence, the northern population has 1,700 m^3/inhabitant/year, and they are not under water stress conditions. Hot spots of water stress are mainly all the countries in southern Europe, North Africa, Lebanon, Israel, Libya and Tunisia. In these areas renewable water (surface and groundwater) does not reach the volume needed for self-maintenance, with less than 1,000 m^3/inhabitant/year and even some areas of less than 500 m^3/inhabitant/year (Milano *et al.*, 2013). In this situation of high demand but low water recharge rate, the solution is to improve and increase infrastructure to store water from rainfall, so that the amount of surface water available for urban and agricultural use is increased (García-Ruiz *et al.*, 2011).

1.3.7 Mexico

The main source of water in Mexico is groundwater, but because of the strong water demand resulting from high population density, aquifers are being seriously overexploited (Chelleri *et al.*, 2015). According to González-Morán *et al.* (1999), 70% of water supply in the Valley of Mexico (45 m^3/s) is obtained from a single aquifer. The problem is that the system is polluted by high tourism pressure on the surface waters that feed the aquifer, as well as high population density in the region. Salt concentrations of up to 25,000 mg/l have been found, in addition to ammonia, nitrate ions, sulfates and iron. This has resulted in many wells having to be closed in order to treat water, which only adds to the pressure on remaining resources. Furthermore, the high frequency of water abstraction provokes a reduction in the water level of the aquifer and can result in ground subsidence, an increasingly common phenomenon in the region. Measured ground subsidence (above the aquifer) can reach a rate of up to 2.2 cm/year, together with the occurrence of fractures of up to 1 m wide and 200 m long. As a result the infrastructure of cities is damaged, i.e., buildings, electricity, water and gas infrastructures, and another consequence of ground subsidence is that fractures increase exposure of groundwater to urban pollution (González-Morán *et al.*, 1999).

Due to high urban density, strong water demand, the environmental impact of excessive water withdrawals and weather causing problems on water reserves, the Mexican government has approved a plan to implement sustainable resource management based on water recharge and water saving, and improving systems of wastewater treatment to promote the use of reclaimed water. All of these measures are aimed at solving the problems of water shortages and protecting the environment from the declining volume of water available. Moreover, given that Mexico City is heavily influenced by extreme weather events that cause strong rains, a way to manage water demand is to capture rainwater to significantly reduce groundwater abstractions. Domestic water demand of the city of Mexico is between 15 and 22 million m^3, of which between 25% and 50% could be covered by rainwater collection. This new management form would be a significant improvement in water management and efficient water use allowing the city to reduce its dependence on groundwater (Chelleri *et al.*, 2015).

1.4 CONCLUSIONS

The current water situation has two main problems. Firstly, there are countries that, despite having favourable rainfall patterns suffer high water extraction due to population patterns and productive sectors. Secondly, there are countries where natural climatic conditions jeopardise the water volume available (as mentioned in Australian water situation analysis). Regardless of the water situation of the countries, all water problems can be solved through suitable water management. According to the examples provided, the worldwide water resources situation is complicated and, in many cases, critical. The problem extends not only to trying to secure urban, agricultural and industrial supply, but also to preserve the environmental conditions of water reservoirs, damaged by high levels of abstractions. There is a widespread tendency to believe that water resources are infinite and that the unsuitable monitoring of available water volumes are mainly responsible for current water stress situations. Ecosystem dynamics are being modified by the lack of water. A glimmer of hope is the use of unconventional water sources that complement conventional water sources. Hence, water uses that consume more water (such as agriculture) can be supplied by seawater desalination or reclaimed water.

Australia provides an example of changes in water management. The Australian government expects to conserve existing resources and guarantee drinking water supply thanks to the inclusion of unconventional sources in the water cycle. This chapter shows that the use of unconventional resources (reuse and desalination) is the best solution for the high water consumption. It is necessary to consider the form of management that we use on water resources, aiming to ensure that all people have access to drinking water, i.e., changing the current management of water resources to achieve sustainable and multidisciplinary water management. Both the demand side (population consumption) and the supply side (volume of water available) need to be considered in order to protect water bodies and guarantee long-term continuity of water resources.

REFERENCES

Australian Government (2013). The need for water reform (Department of the Environment). http://www.environment.gov.au/water/basin-plan/need-water-reform/ Australian Government, Canberra, Australia (accessed 13 July 2015).

Australian Government (2014). The Sustainable Diversion Limit Adjustment Mechanism. Australian Government, Canberra, Australia.

Australian Government (2015). Australian water resources assessment 2012 - Bureau of Meteorology. http://www.bom.gov.au/water/awra/2012/ Australian Government, Canberra, Australia (accessed 13 July 2015).

Australian Government (2016a). Water Information. http://www.bom.gov.au/water/ Australian Government, Canberra, Australia (accessed 1 March 2016)

Australian Government (2016b). Water Storage. http://water.bom.gov.au/waterstorage/awris/ Australian Government, Canberra, Australia (accessed 1 March 2016).

Burn S., Hoang M., Zarzo D., Olewniak F., Campos E., Bolto B. and Barron O. (2015). Desalination techniques – A review of the opportunities for desalination in agriculture. *Desalination*, **364**, 2–16.

Chanan A., Kandasamy J., Vigneswaran S. and Sharma D. (2009). A gradualist approach to address Australia's urban water challenge. *Desalination*, **249**, 1012–1016.

Chelleri L., Schuetze T. and Salvati L. (2015). Integrating resilience with urban sustainability in neglected neighbourhoods: challenges and opportunities of transitioning to decentralized water management in Mexico City. *Habitat International*, **48**, 122–130.

Chowdhury F., Lant C. and Dziegielewski B. (2013). A century of water supply expansion for ten U.S. cities. *Applied Geography*, **45**, 58–76.

Collet L., Ruelland D., Estupina V. B., Dezetter A. and Servat E. (2015). Water supply sustainability and adaptation strategies under anthropogenic and climatic changes of a meso-scale Mediterranean catchment. *Science of The Total Environment*, **536**, 589–602.

Delyannis E. and Belessiotis V. (2010). Desalination: the recent development path. *Desalination*, **264**, 206–213.

Docker B. and Robinson I. (2014). Environmental water management in Australia: experience from the Murray-Darling Basin. *International Journal of Water Resources Development*, **30**, 164–177.

Ejaz Qureshi M., Hanjra M. A. and Ward J. (2013). Impact of water scarcity in Australia on global food security in an era of climate change. *Food Policy*, **38**, 136–145.

Environmental Agency Abu Dhabi (2014). Publications About Abu Dhabi's Environment. http://www.ead.ae/resources/publications/ (accessed 7 July 2015).

European Environmental Agency (2015a). The European Environment – State and Outlook 2015, Synthesis Report.

European Environmental Agency (2015b). Mediterranean Sea Region. http://www.eea.europa.eu/soer-2015/countries/mediterranean (accessed 7 July 2015).

EUROSTAT (2001). Statistics in Focus: Water Management in the Regions of the European Union. KS-NQ-01-013. EUROSTAT, Brussels, Belgium.

FAO (2000). New Dimensions in Water Security: Water, Society and Ecosystem Services in the 21st Century. AGL/MISC/25/2000. FAO, Rome, Italy.

FAO (2002). Crops and Drops: Making the Best Use of Water for Agriculture http://www.fao.org/docrep/005/y3918e/y3918e00.htm (accessed 15 September 2015). FAO, Rome, Italy.

García-Ruiz J. M., López-Moreno J. I., Vicente-Serrano S. M., Lasanta-Martínez T. and Beguería S. (2011). Mediterranean water resources in a global change scenario. *Earth-Science Reviews*, **105**, 121–139.

Ghaffour N., Missimer T. M. and Amy G. L. (2013). Technical review and evaluation of the economics of water desalination: current and future challenges for better water supply sustainability. *Desalination*, **309**, 197–207.

González-Morán T., Rodríguez R. and Cortes S. A. (1999). The Basin of Mexico and its metropolitan area: water abstraction and related environmental problems. *Journal of South American Earth Sciences*, **12**, 607–613.

Gössling S. (2001). The consequences of tourism for sustainable water use on a tropical island: Zanzibar, Tanzania. *Journal of Environmental Management*, **61**, 179–191.

Gössling S., Peeters P., Hall C. M., Ceron J., Dubois G., Lehmann L. V. and Scott D. (2012). Tourism and water use: supply, demand, and security. An international review. *Tourism Management*, **33**, 1–15.

Gu Q., Chen Y., Pody R., Cheng R., Zheng X. and Zhang Z. (2015). Public perception and acceptability toward reclaimed water in Tianjin. *Resources, Conservation and Recycling*, **104 Part A**, 291–299.

Gurung T. R., Stewart R. A., Beal C. D. and Sharma A. K. (2015). Smart meter enabled water end-use demand data: platform for the enhanced infrastructure planning of contemporary urban water supply networks. *Journal of Cleaner Production*, **87**, 642–654.

GWP (2012). Global Water Partnership – The Challenge. http://www.gwp.org/The-Challenge/ (accessed 14 July 2015).

GWSP (2015). Digital Water Atlas. http://atlas.gwsp.org/index.php (accessed 20 September 2015).

Haddad B. M. (2013). A case for an ecological-economic research program for desalination. *Desalination*, **324**, 72–78.

Hernández-Sancho F. and Sala-Garrido R. (2009). Technical efficiency and cost analysis in wastewater treatment processes: a DEA approach. *Desalination*, **249**, 230–234.

Horne J. (2015). Policy issues confronting Australian urban water reuse. *International Journal of Water Resources Development*, **32**(4), 1–17.

Hurlimann A. C. (2009). Water supply in regional Victoria Australia: a review of the water cartage industry and willingness to pay for recycled water. *Resources, Conservation and Recycling*, **53**, 262–268.

Jiang Y. (2015). China's water security: current status, emerging challenges and future prospects. *Environmental Science & Policy*, **54**, 106–125.

Kenny J. F., Barber N. L., Hutson S. S., Linsey K. S., Lovelace J. K. and Maupin M. A. (2009). Estimated use of water in the United States in 2005. *Circular*, **1344**, 52.

MacDonald A. M. and Calow R. C. (2009). Developing groundwater for secure rural water supplies in Africa. *Desalination*, **248**, 546–556.

Marson M. and Savin I. (2015). Ensuring sustainable access to drinking water in sub saharan africa: conflict between financial and social objectives. *World Development*, **76**, 26–39.

McDonnell R. A. (2014). Circulations and transformations of energy and water in Abu Dhabi's hydrosocial cycle. *Geoforum*, **57**, 225–233.

Milano M., Ruelland D., Fernandez S., Dezetter A., Fabre J., Servat E., Fritsch J., Ardoin-Bardin S. and Thivet G. (2013). Current state of Mediterranean water resources and future trends under climatic and anthropogenic changes. *Hydrological Sciences Journal*, **58**, 498–518.

Molinos-Senante M., Hernández-Sancho F. and Sala-Garrido R. (2011). Cost–benefit analysis of water-reuse projects for environmental purposes: a case study for Spanish wastewater treatment plants. *Journal of Environmental Management*, **92**, 3091–3097.

MWR (2012). Ministry of Water Resources the People's Republic of China – Annual Report: 2007–2008. http://www.mwr.gov.cn/english/Publications.html (accessed 10 July 2015).

MWRI (2011). Strategic plan for Ministry of Water Resources. Indian Government, Delhi, India.

Okotto L., Okotto-Okotto J., Price H., Pedley S. and Wright J. (2015). Socio-economic aspects of domestic groundwater consumption, vending and use in Kisumu, Kenya. *Applied Geography*, **58**, 189–197.

Palaniappan M. and Gleick P. (2009). Peak Water (ch. 1). In: The World's Water, vol. **7**, The Biennial Resport of Freshwater Resources (ed.), Pacific Institute, Island Press, Washington DC, USA,.

Radcliffe J. C. (2015). Water recycling in Australia – during and after the drought. *Environ. Sci. Water Res. Tech*, **1**, 554–562.

Richey A. S., Thomas B. F., Lo M., Reager J. T., Famiglietti J. S., Voss K., Swenson S. and Rodell M. (2015). Quantifying renewable groundwater stress with GRACE. *Water Resources Research*, **51**, 5217–5238.

Scarborough H., Sahin O., Porter M. and Stewart R. (2015). Long-term water supply planning in an Australian coastal city: dams or desalination? *Desalination*, **358**, 61–68.

Simons G. W. H., Bastiaanssen W. G. M. and Immerzeel W. W. (2015). Water reuse in river basins with multiple users: a literature review. *Journal of Hydrology*, **522**, 558–571.

U.S. Department of the Interior (2005). The History of Large Federal Dams: Planning, Design, and Construction in the Era of Big Dams. USDI, Washington DC, USA.

USGS (2014). Public Supply Water Use. http://water.usgs.gov/edu/wups.html US Geological Survey, Denver, CO, USA (accessed 12 June 2015).

Walsh M., Wallner G. and Jennings P. (2014). Radioactivity in drinking water supplies in Western Australia. *Journal of Environmental Radioactivity*, **130**, 56–62.

WHO/UN-WATER (2014). UN-water global analysis and assessment of sanitation and drinking-water (GLAAS), 2014 Report: Investing in Water and Sanitation: Increasing Access, Reducing Inequalities. World Health Organization/UN-WATER, Geneva, Switzerland.

WHYMAP (2008). World-wide Hydrogeological Mapping and Assessment Programme - UNESCO. http://www.whymap.org/whymap/EN/Home/whymap_node.html (accessed 22 September 2015).

Wiener M. J., Jafvert C. T. and Nies L. F. (2016). The assessment of water use and reuse through reported data: a US case study. *Science of the Total Environment*, **539**, 70–77.

World Bank (2010). World Development Report 2010: Development and Climate Change. World Bank, New York.

Chapter 2

Urban water demand in a context of scarcity

Mónica Maldonado-Devis and Lledó Castellet-Viciano

Water Economics Group, Department of Applied Economics II, University of Valencia, Avda. Tarongers s/n, 46022 Valencia, Spain

Corresponding author Email: Monica.Maldonado@uv.es

ABSTRACT

In this chapter, an analysis is made of the problem of water scarcity and the importance of urban demand in a context in which water resources are subject to significant pressures. First, the meaning of water scarcity and the most common ways of measuring it are defined. Then, the area's scarcity situation is analysed using two types of indicators, the *Water Stress Indicator plus* and the *drought indices*, both of which are related to water deficit situations. The role of water uses and, in particular, of urban water demand in determining scarcity situations is discussed.

2.1 CONCEPT AND QUANTIFICATION OF WATER SCARCITY

Water is essential for life in both humans and the maintenance of other species and ecosystems. Moreover, it is essential for the economic and social development of different geographical areas (UN, 2003). According to Young and Haveman (1985), there is a "water problem" when water is not in the correct quantity and quality in the right place and at the right time. Although there is a large amount of water on Earth, only a small proportion of it, about 2%, can be used by

Editor(s): Francesc Hernández-Sancho
doi: 10.2166/9781780404684_0023

humans because the remaining 98% is too salty, frozen or just not in the right place at the right time (EC, 2012).

Water is a limited resource, but for the most part is renewable, thanks to the natural cycle that circulates between the earth, the oceans and the atmosphere following a process of evaporation and precipitation. The problems associated with water determine both its supply, that is, the availability of the resource, and its demand. The quantity and quality of available water, together with the evolution and management of the uses of the resource, generate scarcity problems of which an awareness is essential if we want to be able to effectively manage the issue (Watkins, 2006).

The definition and measurement of water scarcity is not unique. While in general terms we refer to the physical scarcity of water as the insufficiency of resources to satisfy the demand (Zetland, 2011), there are also areas of the planet that suffer from other forms of scarcity related to their economic and/or institutional situation. In this way, an area may suffer from an economic shortage of water if, despite having sufficient resources, the economic, human and institutional capacity to allocate it is very limited (World Bank, 2007; Viala, 2008; FAO, 2013). There are geographical areas (countries, regions or sub-regions) in which the inhabitants suffer from water shortages because they lack the necessary infrastructure to transport water from rivers and aquifers, or the existence of public policies that encourage the excessive consumption of certain uses (Watkins, 2006).

When we talk about water problems, it is necessary to bear in mind that we will be facing very different situations in the case of developed and non-developed countries. While less developed countries suffer from shortages related to their economic and institutional situation that prevents people from accessing an essential good, in developed countries there are no problems accessing the resource, and in general terms all people have a sufficient amount of quality, affordable water (AEAS, 2016). The problem faced by developed countries and certain areas within them is rather a situation of physical scarcity caused by the availability of the resource and the evolution of its uses. On the other hand, competition between the different uses of water in many areas of developed countries (especially in cities) exerts strong pressures on water resources, causing greater possibilities of suffering shortages.

According to the Intergovernmental Panel of Climate Change (IPCC), the existing situation surrounding water resources will be aggravated in the future, since climate change is expected to affect its availability and its variability over time, increasing the number of extreme events. These include floods, more frequent periods of drought and storms (IPCC, 2008, 2012, 2014). The IPCC (2014) reports that according to their projections, climate change will reduce both superficial and underground water resources, causing an intensification of competition for the use of water among the different sectors. In this context, urban areas will be particularly affected, and will suffer greater risks from rising temperatures, extreme storms and rainfall, water shortages, droughts and floods (Gosling & Arnell, 2016). In Europe, the size and growth of the urban population and the growing complexity to provide all the necessary services to this population make cities very vulnerable to the effects of climate change, making it necessary to promote adaptation measures that reduce the risks for people (EEA, 2012b).

Problems of water scarcity can be characterised in many ways according to the criteria of supply, use of water or the very definition of scarcity. It is a complex phenomenon that varies depending on the applicable space, time and context (Quiring, 2009). Water scarcity is a local problem that will depend on the measure used and as such will describe different situations depending on the context of the analysis. Different areas will not face the same problems and, therefore, the same solutions cannot be analysed and applied. For this reason, different indicators have been developed over the years in order to assess the vulnerability of water resources (Brown & Matlock, 2011; White, 2012; Pedro-Monzonís *et al.*, 2015). An exhaustive review of the typology of indicators will not be made here, only those that are most commonly used in the scarcity analyses of the main international organisations.

The first ways to measure scarcity are based on the construction of indexes based on human requirements. The simplest and easiest to obtain is known as the Falkenmark Indicator (IF), which is defined as the amount of water available per person per year (Falkenmark *et al.*, 1989). This indicator is constructed with data that are normally accessible at the national level, however it has certain limitations by not taking into account important issues such as the variability of water available within countries, the accessibility of the population to the resource, unconventional water supply sources, reservoirs that increase the natural supply of the resource each year and the differences in the demand of different areas (Rijsberman, 2006). This indicator and its variations have been used frequently in different studies or reports in which the scarcity of water resources is evaluated at a global and European level (EEA, 2010; WWAP, 2012, 2015). According to the aforementioned indicator, the critical value from which a scarcity situation is determined is 1,700 m^3 per person per year.

For its part, the International Water Management Institute (IWMI) is developing another more complex indicator that includes, to the extent of water scarcity, the capacity of users in each country to adapt to the decrease in resource availability using new technologies or other ways to decrease the use of water. For this, the infrastructure of each country, including desalinated and recycled water, is included in the measure of scarcity, and measures its potential for infrastructure development and efficiency improvements (White, 2012). The IWMI classifies countries as either those that suffer from economic shortages and will not be able to meet future demand for water without investment in infrastructure and efficiency improvements, and those that suffer physical shortages and will not be able to cope with future demand, even with that investment.

Another way to approximate the measure of scarcity or water stress is to use indicators that incorporate environmental requirements. The need for a large amount of information in the construction of this measure makes its use very complex at national levels and is more suitable for analysis at a local scale. The Water Poverty Index (WPI) tries to unite the physical availability of water with socioeconomic factors that reflect situations of poverty (Sullivan, 2002). Another example of this type of indicator is the Water Stress Indicator (WSI), which considers the proportion of extractions on the total available resources minus the estimated water requirements for environmental use compared with an indicator without these environmental requirements (Smakhtin *et al.*, 2004).

Measures of the capacity of the planet's water resources to ensure both human and environmental needs take on a new dimension with the introduction of the concept of a "water footprint" that includes both the direct use of water as well as its indirect use (Hoekstra, 2003; Mekonnen & Hoekstra, 2016). The maps shown in Figure 2.1 indicate the degree of scarcity of resources or situations of water stress according to the three indicators mentioned above (IF, IWMI, WSI). These maps make it possible to observe that the use of measures other than water scarcity, depending on whether they consider the capacity to adapt technologically or otherwise, can establish which areas are subject to situations of greater water stress (Vörösmarty *et al.*, 2010).

As can be observed in Map 1 (Figure 2.1), Spain is in a vulnerable situation but is not in danger of shortages, according to the IF, since it exceeds 1,700 m^3 per capita: this is the critical value to establish if an area is subject to water stress. Map 2 presents the situation of physical or economic shortage according to the values of the IWMI indicator. According to this measure, Europe and its regions do not present shortage problems either physically or economically. Map 3 indicates the scarcity situation of the main basins measured through the WSI, including the environmental requirements. It can be observed that the Mediterranean basins present a situation of the overexploitation of resources.

As an alternative to the indicators discussed so far, critical scarcity indices have been developed, which attempt to measure the vulnerability of water resources by incorporating the different demands of each country. This type of index measures scarcity as a proportion of the total annual withdrawals of the total resources available in that year. Although this measure relates the availability of the resource to its demand, it still does not incorporate unconventional offers such as desalination, nor the recycling or reuse of extractions (OECD, 2009; EEA, 2016b). The critical values of these indicators determine if an area is subject to a situation of water stress that is defined as the situation in which the proportion of water extracted from the natural environment to be used in the different activities exceeds a certain threshold from which the pressure on the resource makes it vulnerable (Raskin *et al.*, 1997). In the next section, we analyse the situation of water resources in the Mediterranean Area through measures and indicators using this perspective.

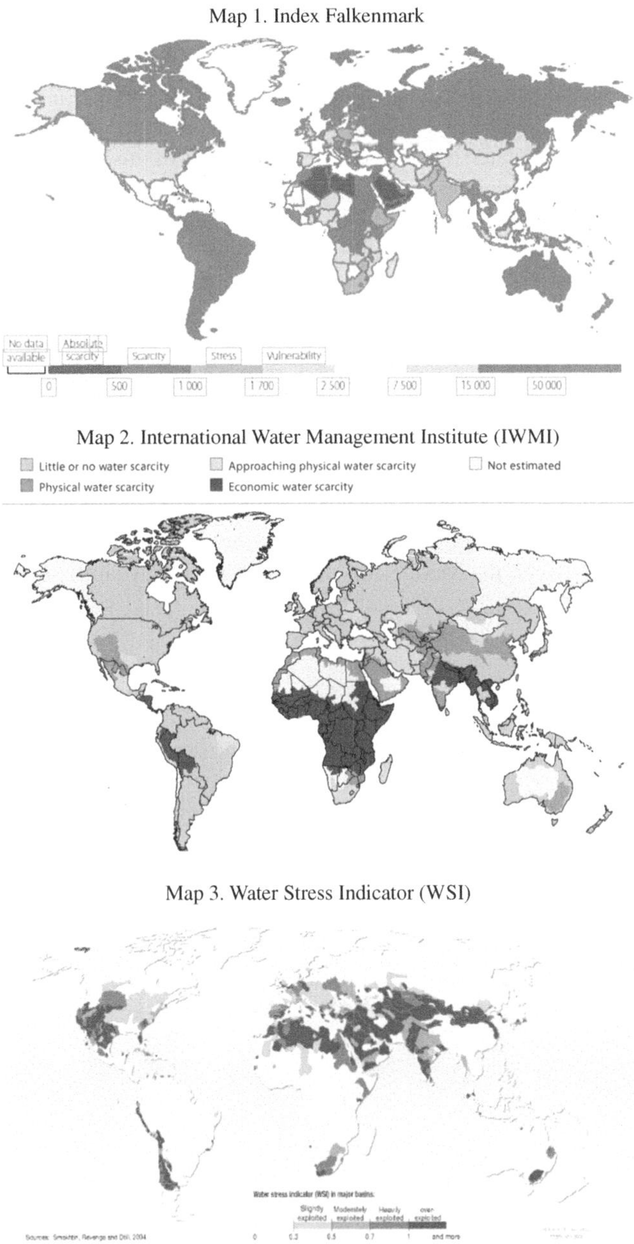

Figure 2.1 Scarcity situation in the world using the following indexes: Map 1 – Index Falkenmark (IF); Map 2 – International Water Management Institute (IWMI); and Map 3 – Water Stress Indicator (WSI). Sources: WWAP (2012, 2015).

2.2 WATER SCARCITY IN THE MEDITERRANEAN REGION

The EU monitors the efficient use of available water resources using a critical shortage ratio called the Water Exploitation Index Plus (WEI+). This indicator analyses how the total use of water exerts pressure on renewable water resources and identifies areas (watersheds or hydrographic sub-basins) with high seasonal uptake in relation to the resources available to them, making them prone to situations of water stress (EEA, 2016b). A WEI+ of more than 20% implies that a water resource is under stress, while one of more than 40% indicates a severe stress and a clearly unsustainable use of resources (Raskin *et al.*, 1997).

If we make the comparison with other regions of the world, water is relatively abundant in Europe, since only 5% of the available resources are taken from the natural environment annually (EEA, 2016b). However, for the rest of the world the availability of water is unevenly distributed geographically and subject to large seasonal differences. The Mediterranean Region, together with the Atlantic, represents 17% of the total renewable water resources extracted from all over Europe. As can be seen in Figure 2.2, which shows the delimitation of the European territory in biogeographical areas, the Iberian Peninsula is divided fundamentally between the Mediterranean region and the Atlantic region.

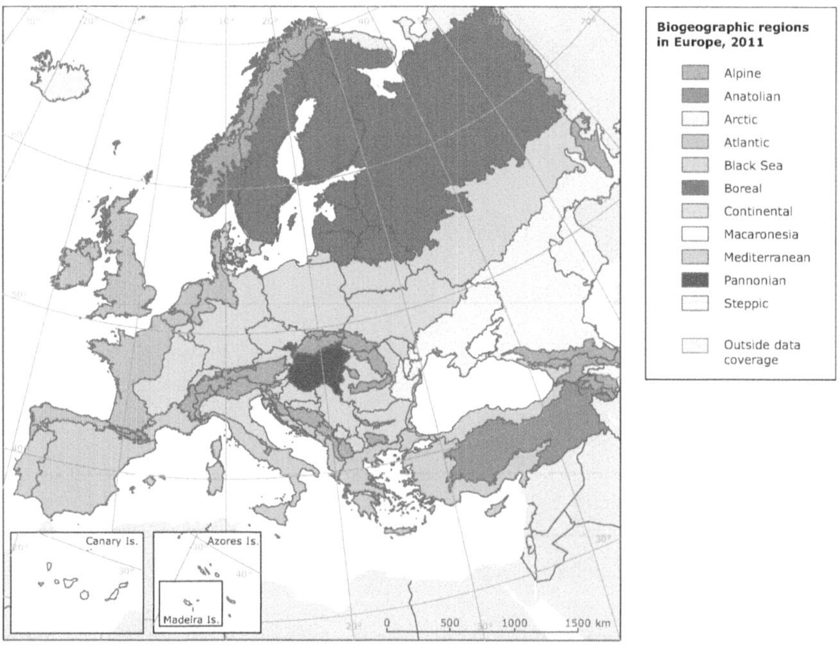

Figure 2.2 Biogeographical delimitation map of the regions in Europe. Source: EEA (2016b).

In the whole of the Mediterranean region, 20% of the population is subject to permanent conditions of water stress; this percentage rises to 53% in the summer months (EEA, 2016b). The average value of the WEI+ in Spain is 34%, although this value does not represent the scarcity situations that occur at sub-regional levels (EEA, 2010). A great diversity of situations can be found in the Iberian Peninsula, caused in part by climatic differences (Hoyos & Artabe, 2016).

The Mediterranean region (like the rest of the European regions) is divided in terms of water resources in Hydrographic Demarcations (Figure 2.3). The city of Valencia is located geographically to the east of the Iberian Peninsula belonging to the European Mediterranean region and, specifically, to the Hydrographic Demarcation of Júcar River. The Valencian Community represents around 50% of the territory covered by the demarcation. The weight of Valencia, the largest city of this region, in terms of pressures on total water resources, is critical.

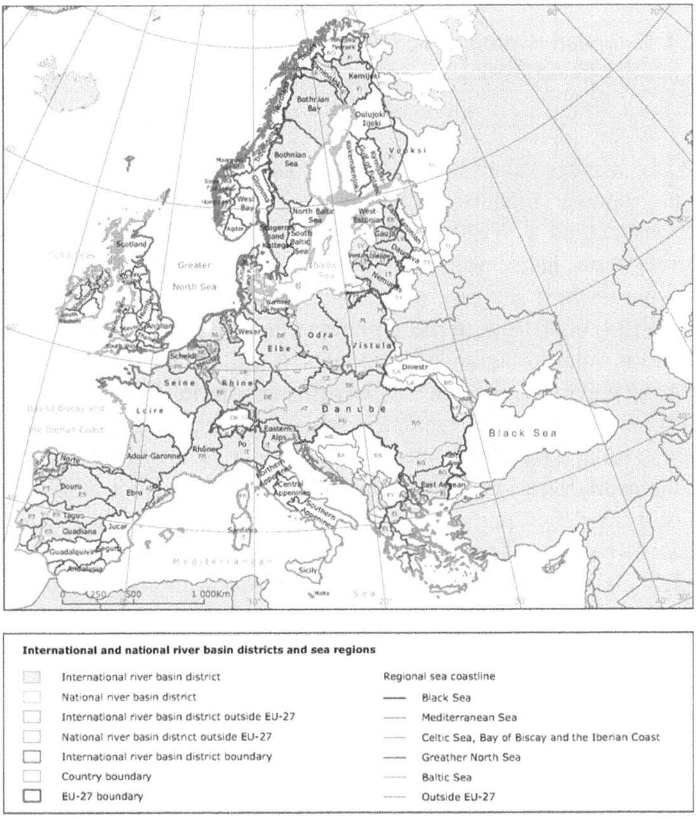

Figure 2.3 Delimitation of Hydrographic Demarcations in Europe. Source: EEA (2012a).

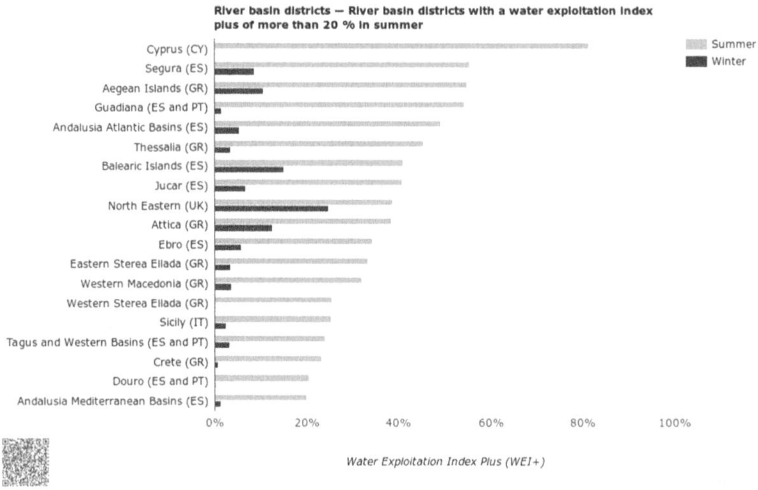

Figure 2.4 European Hydrographic Demarcations with WEI+ greater than 20% in summer. Source: EEA (2016b).

The WEI+ average values of the European Demarcations that exceed a value of 20% in the summer months are shown in Figure 2.4. Eight of the Spanish Demarcations exceed a WEI+ of 20% in this period, and five are subject to severe stress with index values exceeding 40% in the summer. The Júcar Demarcation to which the city of Valencia and its metropolitan area belongs presents a WEI+ of 40.79% in the summer months in which rainfall is lower and the demand for water for agricultural and tourist uses is higher (EEA, 2016b).

As we have previously pointed out, the demarcations must include WEI+ values as part of the water resources assessment process in order to determine those areas most likely to suffer water stress situations. In addition, water balances are made that compare the available resources with the expected demands for all uses so that a situation of water deficit in the demarcation can be evaluated. The last Basin Plan of the Júcar River Basin District established a water deficit of 265 hm^3 per year for the 2015–2021 planning cycle, which means that with the resources available in the demarcation itself it is not possible to meet all the existing water rights and possible future growth of demand with adequate guarantees and comply with thc regime of ecological flows (CHJ, 2015).

As such, the scarcity of water resources that the area presents does not respond to a conjunctural situation that can be limited to the data and water balances made in recent years. The evaluations carried out previously, albeit for territorial areas in some cases, determine a situation in which the available resources are lower than the demands for all uses. The availability of water resources is conditioned by different natural and human factors. One of the determinants of this availability is the climate that, in the area where the city of Valencia is located geographically, is described as typically Mediterranean, with hot summers and mild winters (EC, 2010; CHJ, 2015). Average temperature oscillates between 14 and 16.5°C and the maximum temperatures are registered in the months of July and August. The average annual rainfall in the area is about 500 mm, although there is great spatial and temporal variability. Although the volume of its precipitations places it within what could be called dry Spain, it frequently suffers episodes of heavy localised rainfall in the months of September and October. In this area, it is common for long drought episodes to coexist with large but very irregular rainfall (Rico & Hernández, 2008; Rico *et al.*, 2009).

The concept of drought as scarcity can be defined in different ways (Pedro-Monzonís *et al.*, 2015). As adopted by the EU, a drought situation represents a relevant temporary decrease in average water availability. In order to detect and measure the severity of droughts, different indicators have been developed (Estrela & Vargas, 2012). In Spain, the phenomenon of drought is monitored through the National System of Indicators, whose main function is the implementation of preventative measures. The value of these indicators determines four hydrological states related to the type of actions that are necessary in each case to reduce the impact of the drought on the water resources of a certain area (Pedro-Monzonís *et al.*, 2015).

In Figure 2.5, we can observe the different hydrological states established in the different demarcations of the Iberian Peninsula. According to the data presented so far, the Mediterranean region has high levels of water stress that become severe

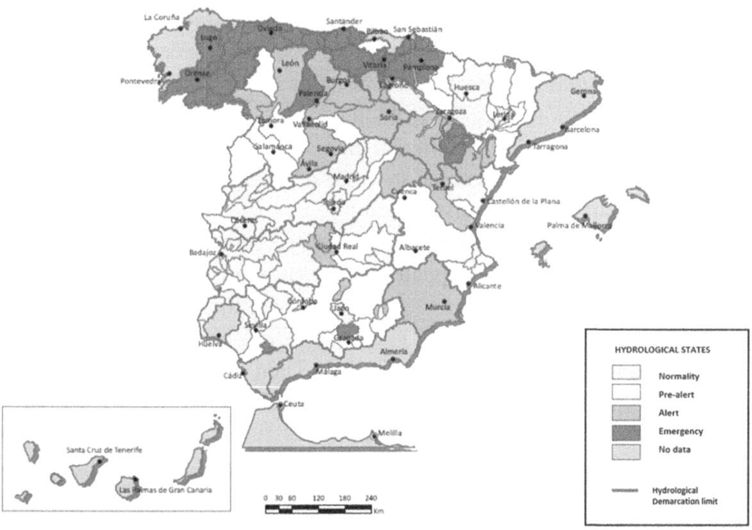

Figure 2.5 Drought indicators, December 2016. Source: www.mapama.es.

during certain periods of the year. The Júcar Demarcation, in particular, presents not only high WEI+ values but also a water deficit situation according to the estimations made in the basin's hydrological plans (CHJ, 2015, 2016, 2017). The prolonged drought phenomena contribute to aggravating the scarcity problems that characterise this geographical area. These scarcity situations are related both to the resources available and to the volume and evolution of its uses and demands.

2.3 TYPOLOGY OF WATER DEMAND: URBAN USES

All economic sectors need water for their activities and, given a limited supply of resources, different uses exert different pressures on renewable water resources. The distribution of these uses of water in different activities is subject to both geographical and seasonal variations, thus generating, in some regions, situations of great pressure on water resources in certain seasons. In general terms, the uses of water are divided into agricultural, industrial and service, and urban and environmental. Agricultural use represents one third of the total resources in Europe (EEA, 2016a), the majority of which is for irrigated crops. According to Eurostat data, in Spain this percentage rises to 67%; at the regional level, the Mediterranean area allocates 75% of the resources for agricultural use (Eurostat, 2017).

The agricultural sector not only uses a large amount of water, but only 50% of it returns to the natural environment (FAO, 2013). From these returns, part filters into the soil together with pesticides and fertilisers, which in some areas leads to serious pollution problems in the aquifers. The problem is especially important in the southern areas of Europe that, due to their arid climate, use 80% of the water resources on agriculture (EEA, 2016a). The industry also needs water in its production processes, although in this case 95% of the water used returns to the environment, but in many cases it does so with pollution conditions that make treatment and purification necessary. Industrial use includes the generation of energy and water used in the cooling of industrial processes. The remainder is divided between electricity generation and refrigeration (17%), and mining, extraction, manufacturing and construction (4%).

The service sector represents a relatively small percentage of the total water used to meet human needs at the European level. However, for some specific areas, activities related to tourism come to represent an important pressure on the environment due to its concentration in very specific periods of the year (EEA, 2009). Service activities, especially those associated with tourism, are also very important in Europe and in certain areas of the Mediterranean region. Tourism in Europe represents 10% of GDP and 11% of the total water consumption corresponds to activities related to the mobility of millions of people to European destinations (EEA, 2016b). These trips and activities are mainly concentrated in large cities (such as London and Paris) or coastal areas (for example, the Mediterranean coast and Mediterranean islands) and are subject to significant variations in the summer months. According to Eurostat data, in 2015 Spain was the main tourist destination for non-residents and the EU member state with the highest net income from tourism (Eurostat, 2016).

Urban uses include diverse activities such as small industries and service companies located in the towns, so-called municipal activities (e.g. irrigation and street cleaning) and supplying households for domestic consumption. The Mediterranean region is the third largest consumer in Europe with a per capita consumption of 133 litres per day, much higher than the European average of 98 litres per capita per day (EEA, 2016b). It is necessary that the rivers maintain their ecological flows in order to preserve the aquatic ecosystems, in the same way that lakes, lagoons or wetlands must maintain the level and quality of their waters. However, the current legislation does not consider environmental demands as a use but as a restriction that must be imposed in the planning of resources. The pressures exerted by the other uses mean that either it is not possible to maintain the ecological flows or that the quality of the water ends up presenting serious problems (EEA, 2010).

According to the European Environment Agency (EEA, 2016b), the total water collection in Europe has reduced between 1990 and 2012. The industrial and agricultural sectors have reduced their water use by more than 20%, while urban use has only reduced by 5% in the same period. As in other analyses carried out in this section, there are also regional differences since there is a decrease in this use in eastern and western Europe, while there has been an increase in the southern regions of Europe. The data from the water supply and sanitation survey show that in Spain, domestic water consumption has increased by 1.5% in the last year (INE, 2016). The intensity and evolution of the different uses may consequently compete and put pressure on the availability of the resource generating situations of water stress. The concentration of agricultural activities is very intensive in the use of water, alongside the higher than average urban use; as such, a tourist activity that strongly diminishes availability during the driest months results in water stress in this region.

Of all the uses of water, urban use represents approximately 12% in Spain, although there are differences between the regions. Of the total water destined for urban uses, the majority (around 70% in Spain) is used for domestic consumption. The average consumption of households in Spain is 132 litres per person per day, although regional differences are certainly relevant. This way, we can find areas in Spain, such as the Community of Castilla-León or the Valencian Community, that consume more than 160 litres/person/day, while others, such as La Rioja or Ceuta and Melilla, consume around 100 litres/person/day (INE, 2016).

The supply of drinking water in Spain is made under a public service regime that has an absolute legal priority in the terms established by the Water Law. This service is also qualified within the scope of the EU as a general interest service in accordance with the Water Framework Directive and, therefore, requires action by public authorities that is not merely complementary, but priority and essential

(Herráez & Vicens, 2010). On the other hand, domestic supply necessitates superior quality and guarantees which greatly impact this use. To satisfy the demand for drinking water, it is necessary to have resources with a high quality of origin, which forces, particularly in large cities, the capture of water from areas far from the supply points. In the same way, the hydraulic infrastructures must be dimensioned to guarantee the supply of water that meets the quality requirements whatever the climatic conditions or water stress that the area presents.

Most of the large Spanish cities distribute water to their inhabitants that is captured outside their municipal boundaries; this usually comes from geographically distant superficial or subterranean sources. This is the case with the city of Valencia, in which a sizeable amount of the water used to supply drinking water comes from the Júcar River and is transferred through the Júcar-Turia canal (Gil & Rico, 2007). Likewise, urban development models tend to generate cities with densely populated nuclei and an increasing number of gardens and pools, and with peripheries predominantly consisting of underutilised secondary housing (Sorribes, 2015). This relationship between water consumption and city urbanisation models has been analysed in the literature (Domene *et al.*, 2005; Domene & Saurí, 2006; Vidal *et al.*, 2011). The incidence of tourism with respect to total water demand can be important at the local level and introduces difficulties due to its high seasonality (OSE, 2008; Trayter, 2009; March & Saurí, 2010; Saurí *et al.*, 2013).

The perspective of climate induced reductions in supply coupled with an increase in demand, aggravated by the effects of population growth, will intensify situations of water stress associated with weather patterns. Climate change threatens to add an additional source of stress to urban water systems (Miller & Yates, 2006; EEA, 2012b).

2.4 CONCLUSIONS

Water scarcity has always been a relevant issue that concerns either developed or developing countries. Not only does water scarcity refer to the quantity of water but also to the quality and the economic and technical resources available to use it. Accordingly, it has been necessary to develop different indicators capable of measuring water scarcity at a global level in every dimension. These indicators cover a wide range of aspects such as socioeconomic features, the amount of water available per person per year, the capability to adapt to shortage of the resource using unconventional water supplies, the efficiency in the use of the resource or even environmental requirements.

Despite the fact that the information used to build the indicators varies from one to each other, all of them share the same aim: determine the level of water stress in different areas of the world. An example of these indicators is the Water Exploitation Index Plus (WEI+) which has been commonly used to assess the level of water stress in different areas of the Mediterranean region, which is subject to high seasonal water stress. However, as important as the assessment of water stress is the analysis the water demand. In the light of this, there is not a unique solution to face a situation of water stress or scarcity. We need to work at a local level and act accordingly to the causes that originate the problem and the characteristics of water demand.

Population growth and its concentration in certain areas and the models of urban development used critically impact water resources in many cases in situations of scarcity. A concentration of both urban and population growth is occurring in territories with scarce resources, such as the Mediterranean coast, thus increasing pressures on a region that is already subject to a certain degree of water stress. The pressure exerted by the supply of drinking water to large urban areas with water deficits or in situations of stress requires the management of these demands that allows for more efficient consumption and contributes to resolving often extreme situations.

REFERENCES

AEAS. (2016). Sostenibilidad social en el sector del agua urbana: Situación y recomendaciones. AEAS, Madrid, Spain (in Spanish).

Brown A. and Matlock M. D. (2011). A review of water scarcity indices and methodologies. White Paper, **106**, 19. The Sustainability Consortium, USA.

CHJ. (2015). Plan hidrológico de la Demarcación hidrográfica del júcar. Ciclo de planificación 2015–2021 Memoria. Anejo 3. Usos y demandas de agua. Available at: http://www.chj.es/eses/medioambiente/planificacionhidrologica/Paginas/PHC-2015-2021-Plan-Hidrologico-cuenca.aspx (in Spanish) (accessed 4 April 2017).

CHJ. (2016). Seguimiento del Plan Hidrológico del Júcar; Ciclo de planificación hidrológica 2019–2015. Available at: http://www.chj.es/es-es/medioambiente/planificacion hidrologica/Documents/Informe-Seguimiento-PHC/SeguimientoPlan2015.pdf (in Spanish) (accessed 2 May 2017).

CHJ. (2017). Informe de Seguimiento de los indicadores de sequía en el ámbito territorial de la Confederación. Oficina de Planificación Hidrológica. Confederación Hidrológica del Júcar. April 2017. Available at: http://www.chj.es/eses/medioambiente/gestions equia/Documents/Informes%20Seguimiento/InformeSequia.pdf (in Spanish) (accessed 25 March 2017).

Domene E. and Saurí D. (2006). Urbanisation and water consumption: influencing factors in the Metropolitan Region of Barcelona. *Urban Studies*, **43**, 1605–1623.

Domene E., Saurí D. and Parés M. (2005). Urbanization and sustainable resource use: the case of garden watering in the metropolitan region of Barcelona. *Urban Geography*, **26**, 520–535.

EC. (2010). Natura 2000 en la Región Mediterránea. Luxemburgo: Oficina de Publicaciones de la UE. European Commission. (In Spanish).

EC. (2012). Comunicación de la Comisión al Parlamento Europeo, al Consejo, al Comité Económico y Social y al Comité de las Regiones titulada: "Plan para salvaguardar los recursos hídricos en Europa". COM 2012/0673 final, European Commission, Bruselas 14 de noviembre. (in Spanish).

EEA. (2009). Briefing n° 3/2008. Impacts of Europe's Changing Climate. European Environment Agency. Official publication of the EU. Luxembourg.
EEA. (2010). The European Environment. Stater and Outlook 2010. Adapting to Climate Change. European Environment Agency, Copenhagen, Denmark.
EEA. (2012a). European water Assessment of status and pressures, EEA Report No 8/2012, European Environment Agency, Official Publication of the EU, Luxembourg.
EEA. (2012b). Urban Adaptation to Climate Change in Europe. European Environment Agency, Official Publication of the EU, Luxembourg.
EEA. (2016a). El agua en la agricultura. European Environment Agency, Copenhagen, Denmark. Available at: http://www.eea.europa.eu/es/articles/el-agua-en-la-agricultura. (in Spanish) (accessed 20 May 2017).
EEA. (2016b). Use of Freshwater Resources. European Environment Agency, Copenhagen, Denmark. Available at: https://www.eea.europa.eu/data-and-maps/indicators/use-of-freshwater-resources (accessed 21 March 2017).
Estrela T. and Vargas E. (2012). Drought management plans in the European Union. The case of Spain. *Water Resources Management*, **26**, 1537–1553.
Eurostat. (2016). Tourism satellite accounts in Europe. Statistical Edition 2015 Reports Luxembourg. EU Publications Office.
Eurostat. (2017). Water Statistics. (Base de Datos en Línea) Available at: http://ec.europa.eu/eurostat/data/database (accessed 2 May 2017).
Falkenmark M., Lundqvist J. and Widstrand C. (1989). Macro-scale water scarcity requires micro-scale approaches. *Natural Resources Forum*, **13**, 258–267.
FAO. (2013). Afrontar la escasez de agua. Un marco de acción para la agricultura y la seguridad alimentaria. FAO, Rome, Italy. (in Italian).
Gil A. and Rico M. (2007). El problema del agua en la Comunidad Valenciana. Fundación de la Comunidad Valenciana Agua y Progreso, Valencia, Spain. (in Spanish).
Gosling S. N. and Arnell N. W. (2016). A global assessment of the impact of climate change on water scarcity. *Climatic Change*, **134**, 371–385.
Herráez P. and Vicens X. M. (2010). Derecho de aguas para entes locales. Bayer Hnos. S.A., Barcelona, Spain (in Spanish).
Hoekstra A. Y. (2003). Virtual water: an introduction. Virtual Water Trade: Proceedings of the International Expert Meeting on Virtual Water Trade. Value of Water Research Report Series 11, pp. 13–23, UNESCO-IHE, Delft, The Netherlands.
Hoyos D. and Artabe A. (2016). Regional differences in the price elasticity of residential water demand in Spain. *Water Resources Management*, **31**, 847–865.
INE. (2016). Estadística sobre el suministro y saneamiento del agua. Nota de prensa datos 2014. Available at: http://www.ine.es/prensa/np992.pdf. (in Spanish) (accessed 15 May 2017).
IPCC. (2008). Climate Change and Water. IPCC, Geneva, Switzerland.
IPCC. (2012). Summary for policy makers. In: Managing the Risks of Extreme Events and Disasters to Advance Climate Change Adaptation, C. B. Field, V. Barros, T. F. Stocker, D. Qin, D. J. Dokken, K. L. Ebi, M. D. Mastrandrea, K. J. Mach, G.-K. Plattner, S. K. Allen, M. Tignor and P. M. Midgley (eds), Cambridge University Press, Cambridge, UK and New York, USA, pp. 1–19.
IPCC. (2014). Climate change 2014: synthesis report. Contribution of working groups I, II and III to the fifth assessment report of the intergovernmental panel on climate change. IPCC, Geneva, Switzerland.

March H. and Saurí D. (2010). The suburbanization of water scarcity in the Barcelona metropolitan region: sociodemographic and urban changes influencing domestic water consumption. *The Professional Geographer*, **62**, 32–45.

Mekonnen M. M. and Hoekstra A. Y. (2016). Four billion people facing severe water scarcity. *Science Advances*, **2**(2), e1500323. doi: 10.1126/sciadv.1500323

Miller K. A. and Yates D. N. (2006). Climate Change and Water Resources: a Primer for Municipal Water Providers. American Water Works Association (AWWA), Denver, CO and IWA Publishing, London, UK.

OECD. (2009). Managing Water for All. An OECD Perspective on Pricing and Financing. OCDE, Paris, France.

OSE. (2008). Agua y sostenibilidad: Funcionalidad de las cuencas. OSE, Madrid, Spain (in Spanish).

Pedro-Monzonís M., Solera A., Ferrer J., Estrela T. and Paredes-Arquiola J. (2015). A review of water scarcity and drought indexes in water resources planning and management. *Journal of Hydrology*, **527**, 482–493.

Quiring S. M. (2009). Developing objective operational definitions for monitoring drought. *Journal of Applied Meteorology and Climatology*, **48**, 1217–1229.

Raskin P., Gleick P., Kirshen P., Pontius G. and Strzepek K. (1997). Water Futures: assessment of Long-range Patterns and Problems. Comprehensive Assessment of the Freshwater Resources of the World. Stockholm Environment Institute, for the United Nations. UN, Stockholm, Sweden.

Rico M. and Hernández M. (2008). Ordenación del territorio, escasez de recursos hídricos, competencia de usos e intensificación de las demandas urbano-turísticas en la comunidad valenciana. Universitat Autonoma de Barcelona, Departament de Geografia, Barcelona, Spain (in Spanish).

Rico M., Olcina J. and Saurí D. (2009). Tourist land use patterns and water demand: evidence from the western Mediterranean. *Land Use Policy*, **26**, 493–501.

Rijsberman F. R. (2006). Water scarcity: fact or fiction? *Agricultural Water Management*, **80**, 5–22.

Saurí D., Olcina J., Vera J., Martín-Vide J., March H., Serra-Llobet A. and Padilla E. (2013). Tourism, climate change and water resources: coastal Mediterranean Spain as an example. In: European Climate Vulnerabilities and Adaptation: A Spatial Planning Perspective, P. Schmidt-Thome and S. Greiving (eds), Wiley Blackwell, New Jersey, pp. 231–252.

Smakhtin V., Revenga C., Döll P., Tharme R., Nackoney J. and Kura Y. (2004). Taking into Account Environmental Water Requirements in Global-scale Water Resources Assessments. IWMI, Columbo, Sri Lanka.

Sorribes J. (2015). Valencia, 1940–2014: Construcción y destrucción de la ciudad. University of Valencia, Valencia, Spain. (in Spanish).

Sullivan C. (2002). Calculating a water poverty index. *World Development*, **30**, 1195–1210.

Trayter J. M. (2009). Urbanismo y Agua. In: El Agua: Estudios Interdisciplinares, M. J. Montoro (Coord.), Institut de l'Aigua, Barcelona, pp. 55–59. (in Spanish).

UN. (2003). Water for People, Water for Life. UNESCO, Paris, France.

Viala E. (2008). Water for food, water for life a comprehensive assessment of water management in agriculture. *Irrigation and Drainage Systems*, **22**, 127–129.

Vidal M., Domene E. and Sauri D. (2011). Changing geographies of water-related consumption: residential swimming pools in suburban Barcelona. *Area*, **43**, 67–75.

Vörösmarty C. J., McIntyre P. B., Gessner M. O., Dudgeon D., Prusevich A., Green P. and Liermann C. R. (2010). Global threats to human water security and river biodiversity. *Nature*, **467**(7315), 555–561.

Watkins K. (2006). Informe sobre desarrollo humano. más allá de la escasez: poder, pobreza y la crisis mundial del agua. PNUD, New York. (in Spanish).

White C. (2012). Global water forum. Understanding water scarcity: definitions and measurements. Available at: http://www.globalwaterforum.org/2012/05/07/understanding-water-scarcity-definitions-and-measurements/ (accessed 16 May 2017).

World Bank. (2007). Making The Most of Scarcity: accountability for Better Water Management Results in the Middle East and North Africa. World Bank, Washingotn, DC, USA.

WWAP. (2012). The United Nations World Water Development. Report 4: Managing water under uncertainty and risk. UNESCO, Paris, France.

WWAP. (2015). World Water Development Report 2015: Water for a Sustainable World. UNESCO, Paris, France.

Young R. A. and Haveman R. H. (1985). Economics of water resources: a survey. In: Handbook of Natural Resource and Energy Economics, A. V. Kneese and J. Sweeney (eds), Elsevier, North Holland, USA, Vol. **2**, pp. 465–529.

Zetland D. (2011). The End of Abundance: Economic Solutions to Water Scarcity, Aguanomics Press, Amsterdam.

Chapter 3

Wastewater treatment and water reuse

Miquel Salgot[1,2]* *and Montserrat Folch*[1,2]

[1]*Soil Science Unit, Department of Biology, Health and Environment, University of Barcelona, Joan XXIII, s/n, 08028 Barcelona, Spain*
[2]*Faculty of Pharmacy, Water Research Institute, University of Barcelona, Joan XXIII, s/n, 08028 Barcelona, Spain*
**Corresponding author Email: salgot@ub.edu*

ABSTRACT

To implement tariffs and regulations on sanitation and wastewater treatment, as well as the disposal or reuse of treated effluents, it is necessary to know the treatment technologies, which one would be best adapted to the present circumstances of the site and finally if the treated wastewater can be disposed of legally or reused in compliance with the rules and regulations in a safe way. Wastewater treatment has been evolving at different paces throughout history, according to the increasing concentration of people in towns and cities. With the increasing pressures on water resources, there are concerns as to how to find new resources capable of helping reach an equilibrium between demand and supply. In this context, one of the main working possibilities is wastewater reclamation and reuse. The basics of wastewater treatment and several of the characteristics of reclamation and reuse are developed in this chapter.

Editor(s): Francesc Hernández-Sancho
doi: 10.2166/9781780404684_0039

3.1 EVOLUTION IN TIME OF WASTEWATER MANAGEMENT

Angelakis and Zheng (2015) refer to the urban water supply and sanitation during the Bronze Age (*ca* 3200–1100 BC) in Crete, the Aegean Islands and Indus valley civilizations. Those civilisations, the first to deal in some way with sanitation, used terracotta pipes for sewage conduction and sewerage and drainage systems. At that time, the contacts of Minoans with Egypt intensified from the period of the first palaces (*ca* 1900–1700 BC). This suggests that a possible flux/exchange of technology related to wastewater in this particular era should exist. The same authors explain that later in the Archaic (*ca* 750–480 BC) and Classical (*ca* 480–336 BC) periods, both historical sources and archaeological excavations provide evidence that wastewater technologies existed and were widespread in Ancient Greece.

The Romans inherited the Greek technologies and further developed them in almost every large city in their Empire. Islamic cultures on the periphery of Europe introduced religious mandates through which high levels of personal hygiene as well as adequate sanitation systems were developed. It is noteworthy that in several cases it was the old Greek and Roman facilities, preserved throughout the centuries, which were used. For thousands of years, untreated wastewater was discharged to rivers or sea, or applied to fields to dispose of it or to maintain soil fertility. If rivers or seawater were not available, or the discharge was considered unsafe, land application was practiced. This kind of application remained active until well into the 20th century.

During the Dark Ages, wastewater management nearly disappeared in the west of Europe, but was maintained in other places of the world. Also in Europe, the use of fields to dispose untreated or partially treated wastewater remained, especially in France and Germany and lasted until the 20th century in some cases. At the beginning of the 20th century, following knowledge gained in the second half of the 19th century, the present concept of sanitation appeared, recovering ancient concepts. In this sense, it is worth noting that activated sludge technology celebrated its centenary during 2014. This technology quickly expanded in the UK and the USA, becoming worldwide the treatment of reference from 1930 onwards. Nowadays, there are a number of technologies based on activated sludge, but other techniques have found their niche in the market.

In parallel, the denominated soft or extensive technologies have evolved, and from the primitive lagooning systems to the more sophisticated wetlands, there are plenty of facilities employing a reduced amount of energy for wastewater treatment. Both types of technique are useful, but in most cases, there is a need to select one of them. In this sense, tools are needed to take decisions on the system to be built. Those tools are reunited in what is called a decision support system (DSS), which considers several aspects to help the DSS model to suggest a few alternatives (Table 3.1).

Table 3.1 Conditions to be included in a decision support system (DSS).

Conditions to Consider		**Remarks**
Wastewater characteristics	Depends on the origin: urban, industrial, agriculture	The loads and its variations exert influence in the final decision
Amount of water to be treated	Every sanitation system has unique water quality characteristics	Usually quantified as inhabitant equivalent (i.e.)
Current rules and regulations	European, national and local rules and regulations	Defines the end quality in a compulsory way
Technology available	Depends on the technological capabilities	The administrations use to be conservative in terms of technology
Landscape integration	Depends on the environmental awareness	Natural systems are better suited for the treatment
Economy	Capacity for operating and maintaining the systems	Theoretically all costs should be included in the price
Social acceptance	Depends on the location and characteristics of the facility	Depends on the communication policies
Centralization/ decentralization	Several towns share a single Wastewater Treatment Plant (WWTP) or each one has their own facility	Can depend on current engineering theories at the building time
Effluent quality	Governed by the regulations, which can even define the accepted technologies	The end use or the disposal can influence the decision
Reuse	The quality is fixed by the legislation	There are some possible uses, with defined qualities for each one
Recycling	Marked by the intended internal uses	Can be successive uses (the last ones are less demanding in terms of quality)
Available surface	The technology should be adapted to the site (surface and form) were the facility will be built	The location of the facility site exerts an influence on the selection of the treatment (e.g. coastal urbanized areas where the facility is inside a building)

Source: Díaz *et al.* (2013).

Water reclamation and reuse is a relatively new concept in many communities of the world, even though it has been practiced for more than 500 years in some regions and is a *de facto* phenomenon in nature and along river basins throughout the world (Salgot, 2017). In many countries, reclaimed water is only a small portion of treated wastewater. With the increase of population shift to cities, reuse is expected to increase in the near future. It is especially practiced in countries where water is scarce, where the arid or semiarid climate causes demand to exceed supply.

There is also the need to define the wastewater treatment system. As a way to copy what natural systems do with the wastewater disposed in them, the extensive treatments for wastewater were developed centuries ago although were not designed in a "professional" way until the 20th century. When deciding which technology should be implemented for a given site, it is necessary to know and compare the characteristics of both types of technology (Table 3.2).

Apart from the treatment technologies, there is a need for additional treatments for specific disposal or for reuse. One of these technologies is disinfection, which is currently applied in several cases and presents different possibilities (Table 3.3).

Salgot *et al.* (2017) deal with several aspects of wastewater treatment, reclamation and reuse in water scarcity conditions. From the last century, due to the need to maintain some control over the amount of water used, water reuse at a large scale will become necessary in order to cope with the demand. Reuse introduces a new stakeholder in wastewater management, with a certain number of implications. Although this practice has been known since the Bronze Age (Angelakis, 2017), it remained in use throughout history (at least in Greek and Roman times) and reached acknowledgement during the 20th century, but it is only since the beginning of the 21st century that it has gained full worldwide knowledge and acceptance, and now reclaimed water is considered as an integral part of water resources.

Table 3.2 Characteristics of extensive and intensive technologies for wastewater treatment.

Intensive Technologies	Extensive Technologies
Energy (net) for mixing and oxygen supply	Soft energy (sun)
Important civil works (concrete, aeration & stirring systems, electric power, pipelines, etc.)	Limited civil works: earth moving, water pipelines
Comparatively reduced surface	Great space occupied
Skilled labour	The staff must know the processes
High-tech "landscape"	Good integration in the landscape
The processes can be quickly influenced	The mechanisms have great inertia
Artificial processes (accelerated using electric energy)	Uses natural components (can be natural processes slightly accelerated)

Table 3.3 Disinfection technologies most used for reclamation/reuse systems.

Technology Type	Variations	Observations
Physical	Ultraviolet radiation (UV)	Several types. Not all are adequate for wastewater
	Gamma radiation	Scarcely used for wastewater treatment and reclamation at small scale
	Membrane-based technologies	Not used after natural systems
Chemical	Chlorination (several types)	Residual disinfection capacity is a must
	Peracetic acid (PAA)	Used to be expensive, now the price has been reduced
	Ozone	Commonly not used after extensive treatments and in small facilities
Other	Additional lagooning systems	Natural UV radiation acts as a disinfectant as well as other mechanisms

3.2 ADVANCES AND IMPROVEMENTS OF WASTEWATER RECLAMATION TECHNOLOGIES

The main concept in relation to wastewater treatment and reuse is the need for safety when treated or reclaimed water goes back to the environment. Existing laws on wastewater treatment define the quality of the water that can be disposed of in the environment, while laws on reuse define the different qualities of the reclaimed water that can be used after the necessary treatment for different purposes.

The classical technologies for wastewater treatment and for reclamation have been used, without much change, for decades. Nevertheless, with the rise of environmental concerns, the requirements for reclaimed water quality and the recent developments for reusing reclaimed water for potable purposes, increasing attention is being paid to treatment and reuse technologies. Coinciding with the need to improve water quality for the indicated purposes, several developments of wastewater treatment and reclamation have been developed during the last decades.

In terms of wastewater treatment, there are several possibilities for development: (a) membrane technologies; (b) natural treatments; and (c) disinfection procedures. In terms of reclamation, technologies must be devoted to the pre-treatment and to the disinfection processes. There are three alternatives for treated wastewater reclamation; namely the use of: (a) secondary treatments capable of obtaining water suitable for reuse; (b) tertiary treatment without disinfection; and (c) full tertiary treatment, including pre-treatment and disinfection. One of the suggestions from several authorities and authors in relation to wastewater reclamation and reuse is to try to improve the image of the whole practice, from the name (wastewater reuse) to a real explanation of the drawbacks and benefits of the practice.

At present, there is a clear tendency to discuss the risks associated with the practice. Although this is a real view of the situation, the concept of risk in relation to water reuse is not a good calling card for a great number of end users, from the farmer who uses the water, to the consumer who uses the product. The drawbacks are related to several items that stretch from the safety of the practice in the long run to the problems in reaching consistently good microbiological and chemical quality. Apart from the quality, safety is also consistent with good practices of reuse and the conviction of the stakeholders that there must be real control from the authorities and if the stakeholders do not guarantee water safety, the supply must be stopped.

There are a lot of studies, papers, pamphlets and, basically, rules and regulations defining the quality of reclaimed water, as well as the suitable precautions and the practices that all related stakeholders are supposed to know and implement. Several countries demand information and communication tasks to further develop this practice and ensure that this is a safe activity. There are several degrees of information, depending on the target audience. Those different levels require different approaches and degrees of information: from the simplest for the farmers and consumers to the most complex to introduce the practice to civil servants and ultimately to politicians. These approaches should be adapted to the needs of every group.

3.3 OBJECTIVES AND REQUIREMENTS OF WASTEWATER REUSE REGULATIONS

Almost all regulations on reuse specify wastewater treatments and among them equivalent treatments, establishing the type of secondary and reclamation (advanced/tertiary) treatment; i.e. the treatment train (California Department of Public Health June, 2014; ISO, 2015; WHO, 2015). The most common secondary treatment (and the most cited) is activated sludge, while sometimes a natural treatment is indicated as acceptable, like lagooning (WHO, 2015). The literature quotations usually refer to equivalent treatments, for example, MBR (membrane bioreactor) or secondary infiltration-percolation as technologies that can be accepted (e.g. Huang *et al.*, 2016; Iglesias, 2016). For tertiary or advanced treatment, coagulation-flocculation, tertiary (maturation) lagoons, membrane technologies or others are also accepted. In terms of disinfection (ISO, 2015), UV radiation, chlorination and ozonization as well as long residence time lagoons are suggested while other disinfection technologies could be accepted or useful (e.g. PAA). Finally, desalination or membrane filtration could be implemented if necessary (Oron *et al.*, 2008).

The effectiveness of treatments is measured by the final quality of the product or by comparison of the initial quality with the final quality. This is known by "validation" of treatments and is nowadays used to prepare the EU regulations on reuse, as in the regulation draft issued by the JRC (2016). An additional indication of quality is given by the effectiveness of the barriers, as indicated by several documents (Alcalde & Gawlik, 2017; JRC, 2016; Marin, 2014; WHO, 1973).

It is worth considering here the present and future regulations, which have been evolving but with a major objective in mind, i.e., safety through analysis of risk (Sharvelle et al., 2017). The evolution of norms and a change in the policy of the EU, having in mind the Circular Economy Concept (EC Communication, 2015), based on a previous Blueprint (EU Communication, 2012), is creating problems of adaptation. There are some similar initiatives, such as the one in South Africa (Ntombela et al., 2015). Because the documents generated (Alcalde & Gawlik, 2014; Amec Foster Wheeler, 2016; JRC, 2016) to comply with the demands of the EU Commission are not fully accepted by the stakeholders, the next months (end of 2018 and beginning of 2019) could be interesting in terms of assessing the final form of the EU regulations.

Right now, reuse is considered a hazardous practice because of the origin of wastewater, mainly because this raw material, which needs to undergo adequate treatment, is polluted by definition. Then, the question arises on "where" the reuse regulations should "start" and if the discharges to the sanitation network (the sewerage) should be controlled (limited and/or analysed) or not. The rationale for this is the inability of common treatments to get rid of a number of pollutants (drugs, pesticides, by-products of disinfection, etc.) which can reach reclaimed waters altered or unaltered, or the entirety of pathogens and their derivatives (e.g. genes which confer resistance to antibiotics).

There are other features that can affect the success of reuse at basin level as described by Simons *et al.* (2015), in terms of the nexus water/energy (Maxwell, 2016), or climate change (Turral *et al.*, 2011). In all cases, reuse is part of the problem and could also be the solution, but one must bear in mind that the impact of the practice is neither absolutely positive nor negative. There are studies which consider that reuse in Europe must be dealt with in a more global approach (Becker, 2012; Simons *et al.*, 2015), or indicating that resources must be drawn from water (Verstraete *et al.*, 2016). Finally, it is important to consider a presentation of Lazarova (Verstraete *et al.*, 2016) making an analysis of the status and recent trends of reuse in Europe and another analysis by Viola *et al.* (2014) studying the water resources in Mediterranean islands, where reclaimed water plays an important role. Table 3.4 summarises different aspects of the rules and regulations.

Rules, regulations and recommendations are keys for the success of wastewater reclamation and reuse practices, and their acceptance by stakeholders is a basic requirement all over the world, especially when other resources are available. There are other approaches, which can help the development of reuse, such as Handbooks of Good Practices (e.g. AEAS, 2012).

Table 3.4 Facts which influence water reuse regulations.

Basic Contents	Minimal Features	Adapted/Existing Tools
Economy and social aspects	End-users acceptation Benefits	Economic calculations (i.e. cost-benefit and other)
Risk assessment/safety	Minimal bureaucracy	LCA (Life Cycle Assessments)
Good reuse practices	Clear and long-standing "rules of the game"	Hazard/Risk assessments (HACCP* and safety approaches)
Policies on formation and information	Public grants/subsidies	Specific support in terms of water quality, agronomy/soil science, environmental impacts, etc.
Communication capacity and understanding of reuse conditions	Help with interpretation of complicated rules	Mid and long-term evaluations

*Hazard Analysis and Critical Control Points.

3.4 POTENTIAL USES OF RECLAIMED WASTEWATER

Reclaimed water can be reused directly (the water is reclaimed and supplied back for several uses), indirectly (after passing through the environment) or for natural water bodies maintenance or augmentation (reclaimed water is returned to natural systems). Almost all rules and regulations are limited to direct reuse. In terms of reuse possibilities (Table 3.5), after the main use for agriculture, groundwater recharge and urban uses should also be mentioned. Industrial uses, landscaping, leisure (e.g. golf) and potable reuse are also practiced. It is usual in reuse to consider, almost always, only the negative impacts on health and the environment; while the positive impacts are scarcely considered, such reduction of extractions from natural water bodies, safety of supply, nutrients for agriculture or less common aspects such as economic and basin management considerations, among other. Several of these are governed by regulations.

There are norms on uses other than agriculture. For example, the EU is considering groundwater recharge distinct from irrigation. In the USA, reuse for potable purposes is gaining momentum and it is possible that regulations on this will appear. There are other places, such as Windhoek, Namibia, where this practice is more than 50 years old (van Rensburg, 2016). There is also a lot of literature on groundwater recharge (Ayuso-Gabella, 2015; Ayuso-Gabella *et al.*, 2011; The Marsol, 2017; Yuan *et al.*, 2016). Industrial uses are also being considered for regulation (Kastner, 2016; Prusakiewicz, 2015) although in some countries these already exist (e.g. RD 1620/2007, Spain). Several types of applications can be grouped, e.g. when reuse is considered as a tool for fighting

Table 3.5 Main reclaimed wastewater uses and related problems (not exhaustive).

Type of Reuse	Problem	In Relation to	End Uses/Comments
Agriculture	Salinity	Soils, Plants	Expensive to be reduced Scarce standards
	Plant illnesses	Plant pathogens	Can affect crop yields
	Toxicity	Accumulation of toxics in matrices	Transmission to the consumers
Urban uses	Environment pollution (urban)	Aerosols, runoff, animal health, etc.	Mainly irrigation, but also cleaning and industrial uses
Industry	Salinity Human health	Scaling Aerosols, pathogens	For refrigeration, cleaning, processes, etc.
Leisure	Human health	Aerosols, ingestion	Irrigation (golf courses, theme parks, etc.), swimming pools
Recovery of water bodies	Environment pollution	Pathogens, toxics, nutrients	Increase the amount of water in the neighbouring bodies
Groundwater recharge	Pollution	Chemicals, physical pollution, microorganisms	Water for several uses including tap water
Potable	Pollution of groundwater	Drinking water safety	Direct reuse, Drinking water

drought episodes. In reality, when droughts occur reuse is potentiated by the authorities, but when the rain returns reuse money disappears.

Economy is one of the basic features in wastewater reclamation and reuse (Winpenny *et al.*, 2010). Although it seems not to have direct connection with rules and regulations, these mark the analytical tasks, the treatments which must/should be used and the interpretation of the results (e.g. risk assessment). Then, there is a direct relationship between legislation, economy of reuse and success of the practice. Although the EU and other instances outwith these countries make emphasise the duty of the end user to cover all expenses, there are more advanced ways to consider cost recovery, because the costs should be shared by all the beneficiaries of reuse systems (Winpenny *et al.*, 2010). This means that not all the costs related to reuse should be attributed to the end user, but also to the other users that obtain benefit from reuse; i.e. the towns that increase their available resources or can reduce the need to store water.

Most regulations establish compulsorily the number of analyses, in terms of parameters and number of samples per time unit. The prices of these analyses must be taken into account considering (Díaz *et al.*, 2013) that: several of the parameters required need highly expensive equipment and are difficult to determine; and as the number of analyses, even the cheaper ones, increases so does the price of the entire analytical procedure. As indicated, reuse is not a practice free of problems, but there are several discussions and several approaches to the rules and regulations depending on the socio-economic characteristics of each country.

3.5 DISCUSSION

A program for the implementation of a wastewater treatment or reclamation project has several major steps, usually: political decision and search for funds; facilities planning (policy); design (type of treatment, location); value engineering, studies on the impacts (either positive or negative) of the facilities; environmental impact evaluation; construction; decision on the type of management; commissioning; building and guarantee period; regular operation and maintenance (O&M); and finally disposal or reuse of treated wastewater. Each step presents its own difficulties, and defining all of them would be a never-ending story or generate a handbook. Nevertheless, several aspects are paramount and have special interest for the end user or the common stakeholder. Those aspects are dealt with more detail in this discussion as well as earlier in this chapter.

Wastewater treatment is paramount to keeping a population safe, i.e. with a good health state. Nevertheless, this can be separated from safety related to tap water. Then, the anthropic water cycle becomes less linear. When including treated wastewater reclamation and reuse, the anthropic water cycle is becoming more of a reality instead of being purely linear. Apart from water supply, wastewater treatment and treated wastewater reclamation and reuse are gaining momentum at present, but are not without economic and environmental concerns. As there is a need to transfer every cost to a final water price, the economic component of the sanitation/reuse systems is important as it signifies a relevant amount of the whole water bill. For this reason, it is important to choose adequate technologies for wastewater treatment and reuse.

The authorities stewarding the anthropic water cycle must optimise the inversions on, and the operation and management of, every facility. The selection of the treatment to be applied to a given site has been, up to now, performed following criteria from engineers, modified or not by politicians. At this level, the engineering and construction companies have their own preferences, based on the available technologies (developed and owned by the company) and the relative operation and maintenance costs. This sometimes has lead to the implementation of old, but patented, technologies and to the clear neglect of the soft technologies. Another explanation could be the difficulties, and operation & maintenance problems, of treatment systems which have just arrived on the market. Those new technologies are not well known by the staff of facilities, which creates problems and faulty functioning of the wastewater treatment plants.

Part of this issue can also be attributed to the centralisation of the decisions, instead of leaving the selection to local authorities more sensible of the day-to-day costs of wastewater treatment. The decision becomes more complicated if reclamation and reuse enter the scenario. It is necessary to couple the treatment system with a new reclamation facility. This is not easy, because the first (wastewater treatment) is governed by disposal rules and regulations, while reuse has its own specific regulations. It can also happen that the manager of the first facility is not the same as for the second and even that the end user receiving the treated wastewater is the owner of the reclamation facility.

In order to "technify" such decisions – on which treatment to use in every case – or at least define the useful technologies and support the decision makers, there is a tool called Decision Support System (DSS), capable of offering a number of technologies adapted to a situation, and discard others that are not useful in a specific case. Once the decision has been taken on every step of the treatment, it can happen that three different companies manage the three steps: treatment, reclamation and reuse. Of course, this can create problems on the attribution of failures or bad water quality obtained.

Further on, although it seems that reclamation and reuse should be a practice with the support of the stakeholders, there are several barriers/limitations for its real development (among others): financing, the most important problem when planning reuse since it is difficult to gain full support from the authorities or contributions of end users; rules and regulations on reuse are not fully adequate and sometimes are not reasonable. This situation causes delays and public opinion problems, in addition to quality concerns; there can exist some unreasonable institutional policies; education and information on reuse requires further implementation to almost all stakeholders; and some stakeholders do not rely on the water authorities. This is due to the lack of clear explanations or because people expect to be dispossessed of their rights to "old" resources, when changes of water source are demanded/executed.

There are also technical problems, but at present these can be solved, given sufficient economic resources. At present (2018), in Europe rules and regulations are "boiling" as the Commission is trying to implement the risk approach in reuse regulations, but with unclear and not technically correct interpretations. These regulations in this case will be limited to reuse for agriculture. In addition to the classical California and EPA (USA) and Israel reuse implementers, there are brand new areas leading reuse, such as Australia, Namibia and several Mediterranean countries. In a number of occasions, reuse starts because of a structural lack of water and the users start to reuse lawlessly.

The main user of reclaimed water is agriculture. Note that at present, there are different ways to manage crops, from the classical way (with intermediaries/ brokers) to the proximity markets, bio- or ecological products, and also "auto-production" (with or without brokers). This dispersion of the market, with diverse distribution channels, is complicating the control of reclaimed water in agriculture. This control is basic to ensure the safety of the consumed crop.

Groundwater recharge is now growing, at the same time that new approaches of the practice are being implemented. Managed Aquifer Recharge (MAR) is a paradigm in this sense, overcoming classical recharge procedures. It appears that passage through an aquifer is a way to reduce the risks, since it is a good barrier for pathogens and several chemicals and at the same time, the aquifer acts as a reservoir but without the problem of evaporation.

Urban and leisure uses for reclaimed water (parks and gardens, golf courses, theme parks, etc.) present special characteristics, depending on the possibilities of contact with the users or other persons as well as interactions with the environment. These uses are regulated by few countries, and present some risk derived from the application site. Industries have been relying on reclaimed water for a long time, especially the bigger users. This practice, extended for decades in the States and other countries, has not been fully accepted in Europe. The practice requires large investment but guarantees the industry a constant, safe supply in terms of quality and quantity.

Water bodies' recovery has been a usual practice, since untreated wastewater has been disposed of in water bodies for millennia. During the 19th and 20th centuries, with the development of knowledge on microbiology, chemistry and environmental engineering, it become apparent that it was necessary to manage differently the elimination of wastewater (treated or not). In sensitive, protected areas, it will be compulsory to fully treat wastewater (advanced or tertiary treatments) before it reaches the environment.

The case of reuse for potable water is a paradigm of certain inconsistencies. While there were plenty of resources available, potable reuse was considered practically heretical, with the exception of a few places, like Windhoek, the capital of Namibia (van Rensburg, 2016). When in California and other places water became really scarce and no more resources were available for supplying towns, the practice seemed acceptable in the US and other countries in the world. Of course, there is a risk associated to potable reuse, which could be greater than for other uses. Nevertheless, present technologies and the implementation of several barriers can lead to an acceptable risk in this practice.

3.6 CONCLUSIONS

Wastewater is a tool for improving water resources management at basin or national level, even if interbasin transferences exist. In the future, water distribution could be planned including non-conventional resources through centralized management, such as in Cyprus. Prices for reclaimed water are usually fixed depending on the origin of the water, and can increase if analytical prices and other expenses are included.

Agricultural reuse is still in first place in terms of amount of water employed, but other alternatives are gaining importance, like urban or industrial uses. Several agricultural developments at present (e.g. proximity agriculture, permaculture, urban roofs, etc.) are also using water from different origins. Reuse in golf courses pushed economic development in golf destinations, such as in the south of the Iberian Peninsula. Groundwater recharge is increasingly studied, and renamed with several terms, for example, Managed Aquifer Recharge (MAR).

Legionella associated risks, related to aerosol formation, are considered to be an emerging risk in reuse. Reuse must be complemented by other resources in global management systems. Potable reuse is emerging as the real ultimate solution when droughts or structural water scarcity make it competitive (e.g. with the reverse osmosis technologies), or when energy is also scarce or not available, for urban water supply.

REFERENCES

AEAS. (2012). Handbook of Good Reuse Practices. Commission V, AEAS. Madrid, Spain.

Alcalde L. and Gawlik B. M. (2014). Water Reuse in Europe – Relevant guidelines, needs for and barriers to innovation. Publications office of the European Union. EU, Luxembourg. (to be incorporated).

Alcalde L. and Gawlik B. M. (2017). Water Reuse in Europe. Relevant Guidelines, Needs for and Barriers to Innovation. A Synoptic Overview, JRC Science and Policy Reports. European Commission, Luxembourg.

Amec Foster Wheeler. (2016). EU-Level Instruments on Water Reuse, Final report to support the Commission's Impact Assessment, London, UK.

Angelakis A. N. (2017). Hydro-technologies in the Minoan area. *Water Supply*, **17**(4), 1106–1120.

Angelakis A. N. and Zheng X. Y. (2015). Evolution of water supply, sanitation, wastewater and stormwater technologies globally. *Water*, **7**, 455–463.

Ayuso-Gabella N. (2015). Risk Assessment and Risk Management in Managed Aquifer Recharge and Recycled Water Reuse: The Case of Sabadell. 2015 PhD thesis, University of Barcelona, Spain.

Ayuso-Gabella N., Page D., Masciopinto C., Aharoni A., Salgot M. and Wintgens T. (2011). Quantifying the effect of Managed Aquifer Recharge on the microbiological human health of irrigating crops with recycled water. *Agricultural Water Management*, **99**, 93–102.

Becker N. (2012). Impact of framework conditions and steering instruments on water recycling in Europe. PhD Thesis, University of Applied Sciences and Arts Northwestern Switzerland, Muttenz, Switzerland.

California Department of Public Health. (2014). Regulations related to Recycled Water, June 2014 (Revisions effective on 6/18/2014) Sacramento, CA, USA. Title 22. California Department of Public Health, Sacramento, CA.

Díaz A., Folch M. and Salgot M. (2013). Wastewater reuse: the economic cost of complying regulations. 3rd International Conference on Water Economics, Statistics and Finance, Marbella, Spain.

EU Communication. (2012). A Blueprint to Safeguard Europe's Water Resources. COM (2012)673. EU Commission, Brussels, Belgium.

EC Communication. (2015). Closing the loop – An EU action plan for the Circular Economy. COM (2015)614. EU Commission, Brussels, Belgium.

Huang C., Zeng P., Yang S., Shao Y. and Lui Y. (2016) Water Reclamation and Reuse. *Water Environment Research*, **88**, 1261–1278.

Iglesias R. (2016). La reutilización de efluentes depurados en España: Retrospectiva, desarrollo del marco normativo, estudio de las tecnologías de regeneración frente a los biorreactores de membrana y sus costes en función del uso. 2016. PhD thesis, Universidad Politécnica de Madrid, Spain. (In Spanish).

ISO. (2015). International Standard Guidelines for Treated Wastewater use for Irrigation Projects – 4 Parts: ISO 16075-{1 to 3}:2015 and ISO 16075-{4}:2016. International Standards Organization, Geneva, Switzerland.

JRC. (2016). Draft V 3.2. Development of minimum quality requirements for water reuse in agricultural irrigation and aquifer recharge. Unpublished paper.

Kastner M. J. (2016). Workshop on 'Water reuse policies for direct and non-direct potable and industrial reuses', 15–16 June 2015, Singapore. *International Journal of Water Resources Development*, **32**, 167–170. DOI: 10.1080/07900627.2015.1080154.

Marin R. (2014). Control de calidad en las aguas residuales y regeneradas: parámetros a controlar en función de las normativas aplicables y nuevas tendencias (Quality control for wastewater and reclaimed water: control parameters in relation with applicable rules and new tendencies). *Tecnoaqua*, **5**, 50–63. (In Spanish)

Maxwell S. (2016). The water-energy nexus in water and wastewater treatment. *Journal AWWA*, **198**, 20–24.

Ntombela C., Funke N., Meissner R., Steyn M. and Masangane W. (2015). A critical look at South Africa Green Drop Programme. *Water SA*, **42**, 703–710.

Oron G., Gillerman L., Buriakovsky N., Bick A., Gargir M., Dolan Y., Manor Y., Katz L. and Hagin J. (2008). Membrane technology for advanced wastewater reclamation for sustainable agriculture production. *Desalination*, **218**, 170–180.

Prusakiewicz J. (2015). Recycled water – alternative supply for power plants. World Water: Water Reuse & Desalination, Autumn, 16–18.

Royal Decree. (2007). 1620/2007, Spain. Royal Decree on reuse of treated wastewater. Boletín Oficial del estado. December 07, 2017.

Salgot M. (2017). État des lieux, perspectives de la réutilisation. Séminaire sur la réutilisaton des eaux usées traitées. Wilaya de l'Oriental du Maroc, Oujda, Morocco. (In French).

Salgot M., Oron G., Cirelli G. L., Dalezios N. R., Díaz A. and Angelakis A. N. (2017). Criteria for wastewater treatment and reuse under water scarcity. In: Handbook of Drought and Water Scarcity: Environmental Impacts and Analysis of Drought and Water Scarcity, T. C. Crusberg and S. Eslamian (eds), Chapter 14, CRC Press, Boca Raton, FL, pp. 263–228. DOI: 10.1201/ 9781315226781-15.

Sharvelle S., Ashbolt N., Clerico E., Hultquist R., Levernz H. and Olivieri A. (2017). Final report. Risk-Based Framework for the Development of Public Health Guidance for Decentralized Non-Potable Water Systems. National Water Research Institute for the Water Environment & Reuse Foundation Alexandria, VA, USA.

Simons G. W. H., Bastiaanssen W. G. M. and Immerzeel W. W. (2015). Water reuse in river basins with multiple users: A literature review. *Journal of Hydrology*, **522**, 558–571.

The MARSOL Project. (2017). Demonstrating Managed Aquifer Recharge as a Solution to Water Scarcity & Drought. http://www.marsol.eu (accessed online 24th July 2017).

Turral H., Burke J. and Faurès J. M. (2011). Climate Change, Water and Food Security, FAO Water Reports 36, FAO, Rome, Italy.

Van Rensburg S. 2016. Overcoming global water reuse barriers: the Windhoek experience. *International Journal of Water Resources Development*, **32**, 622–636.

Verstraete W., Cornel P., Pikaar I. and Li H. (2016). Resource recovery from water. In: Global Trends & Challenges in Water Science, Research and Management, 2nd ed. IWA Publishing, London, UK.

Viola F., Sapiano M., Schembri M., Brincat C., Lopez A., Toscano A., Diamadopoulos E., Charalambous B., Molle B., Zoumadakis M., Torrens A, Gallinas M. and Noto M. T. 2014. The state of water resources in major Mediterranean Islands. *Water Resources*, **41**, 639–648.

Winpenny J., Heinz I., Koo-Oshima S., Salgot M., Collado J., Hernandez F. and Torricelli R. (2010). The Wealth of Waste. The Economics of Wastewater Use in Agriculture. FAO Water Reports 35, FAO, Rome, Italy.

WHO. (1973). Reuse of Treated Wastewaters: Methods of Wastewater Treatment and Health Safeguard. Report of a WHO Meeting of Experts. Technical Report Series No. 17. World Health Organization, Geneva, Switzerland.

WHO. (2015). Sanitation Safety Planning. Manual for Safe Use and Disposal of Wastewater, Greywater and Excreta. World Health Organization, Geneva, Switzerland.

Yuan J., Van Dyke M. I. and Huck P. M. (2016). Water reuse through managed aquifer recharge (MAR): assessment of regulations/guidelines and case studies. *Water Quality Research Journal*, **51**(4), 357–376. DOI: 10.21667wqrjc.2016.022.

Chapter 4

Instruments of urban water demand management

*Monica Maldonado-Devis and Francesc Hernández-Sancho**

Water Economics Group, Department of Applied Economics II, University of Valencia, Avda. Tarongers s/n, 46022 Valencia, Spain

**Corresponding author Email: Francesc.Hernandez@uv.es*

ABSTRACT

In the context of domestic water demand management, the most used instrument is pricing. It is considered that price mechanisms constitute an efficient information system that determines the decision making by producers and consumers. However, it is increasingly accepted that economic instruments should be combined with other types of regulatory policies. In the process to determine the total price of water, it is necessary to identify what objectives have been set, what type of price structure best fits these objectives and, finally, what form or legal nature each payment will take. Obviously, different price structures will send different signals and the design of a specific water pricing structure will depend on the established objectives. In this chapter the urban water prices for drinking water supply in some Spanish cities are analysed.

4.1 ECONOMIC INSTRUMENTS FOR URBAN WATER MANAGEMENT

The holistic approach to Integrated Water Resources Management (IWRM) has been accepted internationally as the best way to manage increasingly limited

 Water Consumption, Tariffs and Regulation
Editor(s): Francesc Hernández-Sancho
doi: 10.2166/9781780404684_0055

water resources and competing demands. In this context, demand management plays a predominant role in contrasting the supply approach when it comes to achieving the objectives of efficiency, environmental sustainability and equity (García-Rubio *et al.*, 2015). This change in the orientation of water policies is reflected at the European level with the approval of the Water Framework Directive and the subsequent developments in this area that attached great importance to economic instruments when managing water resources (EA, 2000; EEA, 2010).

Traditionally, public intervention in environmental policy can be developed through normative, mandate and control instruments, economic and market instruments, and voluntary instruments (Labandeira *et al.*, 2007). The normative instruments are regulations that establish restrictions when carrying out certain behaviours that are considered harmful to the environment. The regulations must be adhered to, and control systems are in place in the event of non-compliance; these carry heavy penalties. The least flexible way to introduce these instruments is to establish restrictions or standards. These types of instruments are considered very effective and usually have low management costs, although they are less efficient than market instruments.

Faced with these types of environmental policy instruments, we identified economic instruments that are defined as tools designed to be implemented with the purpose of adapting individual decisions to collective objectives. On the whole, economic instruments can use price or quantity mechanisms. In the former, the regulator modifies the price through taxes or subsidies and the agents react to that change by modifying the amounts. In the case of quantity mechanisms, the regulator intervenes by setting the optimal amount and the agents, through their actions, will modify the market price.

If we focus on water policy, we can distinguish three basic forms that these instruments can adopt (Strosser *et al.*, 2013; Zetland *et al.*, 2013): (1) use of water prices to generate incentives through tariffs or taxes; (2) mechanisms of quantities through the markets of rights of use or pollution; and (3) cooperative mechanisms based on the voluntary adoption of new practices that reduce pressures on resources. The latter can be adopted without any payment (voluntary instruments) or through subsidies (price instruments).

Price mechanisms have received the most attention in the field of water resources in general and in urban areas in particular, taking into account that Article 9 of the Water Framework Directive (WFD) establishes the need to use prices as an instrument to recover costs, as in their role as incentives to improve efficiency, thereby contributing to the achievement of the environmental objective set out in the directive (EA, 2000). However, it is increasingly considered that economic instruments should be used in combination with other types of regulatory policies, in terms of standards or the imposition of fines for non-compliance with leakage reduction rates, or for raising awareness of users to the adoption of new technologies and consumption habits (Collins *et al.*, 2009; Labandeira *et al.*, 2007).

In terms of the field of domestic water demand management, the recent literature distinguishes between price and "non-price" instruments (García Valiñas *et al.*, 2015). Among the so-called non-price instruments, such regulatory measures as restrictions on the use of water, information campaigns that affect consumption habits and incentives to adopt new water-saving technologies can be found. However, the most widely used instrument in the management of domestic water demand is still pricing: this is a powerful instrument that constitutes an efficient information system to determine the decisions taken by producers and consumers (OECD, 2010).

4.2 WATER PRICES FOR DOMESTIC CONSUMPTION AS A MANAGEMENT INSTRUMENT

The price is the total amount that a consumer pays on their bill for a consumed unit of water. According to the glossary of terms of the EEA, the price of water is defined as the monetary value at which water can be bought or sold (EEA, 2012). The total price of water for domestic uses must include all services associated with the urban cycle in compliance with the principle of cost recovery included in the WFD. The water user must pay not only for the supply service but also for the collection, distribution and sanitation of water.

Prices can be used to encourage the efficient use of the resource, but they can also be used to promote the internalisation of environmental and social costs, to raise revenues, to cover the costs of the service and infrastructures or to achieve equity objectives (Massarutto, 2007). The complexity when setting a price for domestic water consumption is due, in many cases, to its need to meet multiple objectives (OECD, 2003, 2009, 2010). Although it is possible to achieve more than one objective at the same time, there are often conflicts between them that make it advisable to use other instruments outside the tariff.

To determine the total price of water, it is necessary to decide what objectives have been set, what type of price structure best fits these objectives and, finally, what form or legal nature each payment will take. Different price structures can send different signals; the design of a certain water pricing structure will depend on what the underlying objectives are.

To comply with the aforementioned objectives, the competent administrations in each phase of the cycle must decide on the design of the price of the service provided to the consumer. A price structure is a set of rules or procedures that determine the conditions of service and payments for various categories of water users (AEAS, 2014).

The different components of water prices may include fixed payments, payments independent of the volume consumed and/or volumetric payments depending on the amount consumed by the user. Fixed payments can be uniform or differentiated between consumers with different characteristics (socioeconomic, meter calibre, seasonal or dwelling characteristics). In the same way, volumetric payments can also be established in various ways depending on the underlying objectives of the pricing structure.

The first classification distinguishes between those cases in which all the units consumed have the same price (uniform), and in which different prices are established for different units (non-uniform). Non-uniform prices can be segregated into two types depending on whether they differ by blocks of consumption or if they differ by the total volume of consumption. In block rates, all the units consumed in a certain section have the same price and the units consumed in different sections have different prices. If the prices applied to each unit are higher, the greater the consumption segment is, the higher the price per growing block (increasing block tariff – IBT) will be, in which the first units consumed are paid at a lower price than the following units once the block limit has been exceeded. Similarly, if the applied price is lower, the greater the consumption section, the lower the price of the decreasing blocks. In this case, the first units are paid at a higher price than the following units once the block limit has been exceeded. This type of pricing is not usually used in water supply services.

The further type of non-uniform pricing is the so-called volume differentiated tariff (increasing rate tariffs – IRT), also known as tariffs with a disappearing first block (Komives *et al.*, 2005). These are progressive rates in cases of water usage, although in this kind of tariff the user pays the same price for all units of water consumed, but the variable price paid per unit increases with use. This price increase is determined by blocks, but the total amount of water consumed is paid at the price of the last block reached. Like the fixed charges, the volumetric payments can also be adjusted so that the prices of each block or the limits of the blocks are different for specific groups of consumers (family size or user income).

The types of prices described above, summarised in Table 4.1, are combined in the practical application, thus resulting in a multitude of different structures. Each part of the urban cycle, as well as the competence of different administrations, will determine the form of payment for the service provided. The structure of the final price will correspond to the aggregation of all the payments fixed in each phase of the cycle. It is very common for domestic water prices to be "in two parts" combining a fixed charge with a volumetric rate. Depending on whether a uniform or differentiated fixed charge is applied and which type of volumetric fee is applicable, this may result in very different tariff structures and, therefore, also achicvc diffcrent objectives.

The generalised application of a fixed charge and its increasing weight on the total of the invoices adheres to the objective of thc providers of the financial sustainability service (OECD, 2009, 2010). IBT structures are widespread because they can be designed in such a way that less consumption means a lower price thus guaranteeing the minimum affordable consumption for lower income households; on the other hand, higher consumption pays a higher price helping to avoid excessive consumption. However, evidence has been found that at least

Table 4.1 Classification of types of water-pricing structures.

Payment Type	Sub-type	Description	
Fixed charges	Differentiated	A fixed amount that may be different for different users or groups is paid. The differentiation can be made by meter calibre, season, time discrimination, family size or income levels.	
	Not differentiated	An equal fixed amount is paid by all users.	
Volumetric payments	Uniform	The payment depends on the units consumed but the same price is paid for all of them.	
	Non-uniform	Growing blocks (increasing block tariff – IBT)	The payment depends on the consumption, but a higher price per unit is paid for the higher consumption levels. The units are paid at different prices.
		Decreasing blocks	The payment depends on the consumption, but a lower price per unit is paid for the higher the consumption level is. The units are paid at different prices.
		Differentiated by volume (increasing rate tariffs – IRT)	The payment depends on the consumption, but a higher price per unit is paid for the higher the consumption level. The units are paid at the same price (to the last consumption block)

some of these objectives have not been met after the application of this type of structure, since it either does not generate the necessary incentives for the efficient use of the resource or it generates adverse redistributive effects.

When prices present non-linear structures, a difference appears between the average price and the marginal price paid for a unit of water. The design of a non-uniform price must contain at least three parameters for each water user: the number of blocks, the amount of water in each block and the price per additional unit in each block.

In addition, what subscribers pay to a public service of drinking water supply is due to figures of a different legal nature. As Jiménez (2012) establishes, in practice, users pay taxes, public prices and other prices, which are often referred to as tariffs. Various types of taxes are used in the system of taxation on water, whether taxes, fees or special contributions, which in many cases are referred to as fees (Sartorio, 2009). Finally, it will be the combination of the objectives assigned to the price together with the decision of the type of structure being proposed for each service provided and the adopted legal nature that sets out a total water price for domestic use.

4.3 URBAN WATER TARIFFS IN SPANISH CITIES

In Spain, where a significant part of the territory is subject to situations of water stress, price policies in general and in cities in particular have acquired great relevance in recent years (Collins *et al.*, 2009; Fuentes, 2011; García-Rubio *et al.*, 2015). In general terms, pricing structures in Spain are in two parts. The fixed part depends either on the size of the meter (the most common type) or on the type of housing, or is associated with a minimum consumption. Only 2.9% of the population is not charged with a fixed part. The most applicable variable part is a growing block rate while the uniform rates are used sparingly (only 5.2% of the population has a uniform rate), whilst IRTs are virtually non-existent (Arbués & Barberán, 2012).

However, given that the supply of drinking water is a municipal responsibility, in Spain there exists a great diversity of price structures; it could be said that there is a different water price for each of the Spanish municipalities. In some cases, the differences are even more pronounced since the supply rates are not established per household (which is more usual) but *per capita* whenever the municipal census is accredited.

As an example of the great differences that can occur in payments for water in Spain, the rates of four municipalities are analysed in Table 4.2. The payments for potable water supply in the municipalities of Seville, Malaga, Barcelona and León in 2016 have been analysed. Data on the information offered by the different water managers have been obtained from the applicable websites. The types of fixed and variable payments are included in the analysis, as well as the type of bonuses or grants that apply in these municipalities. Bonuses in

Table 4.2 Water tariffs in different Spanish cities 2016.

City	Fixed Part	Variable Part	Bonuses
Sevilla	Fixed quota: According to the diameter of the water meter	IBT 3 blocks: Limit blocks per capita or per house.	Bonus responsible consumption (1 block only for accredited inhabitants)
Malaga	Fixed quota: According to the diameter of the water meter	IBT 4 blocks: Limit blocks per capita or per dwelling. Price blocks different for per capita and for housing	Social fund
Barcelona	Fixed fee: depends on the type of housing. Flow installed. Water points	IBT 5 blocks: Consumption limits per dwelling	Change in the boundaries of the blocks in houses more than 3 people. Social tariff. Solidarity fund.
Leon	Fixed fee: minimum consumption 20 m^3/quarter	IRT 3 blocks: Limits for housing. All consumption is paid at the price of the last block reached.	Exemptions and bonuses for retired, large family, low income, unemployed.

payments for water may be related to the volume of consumption or to the socioeconomic characteristics of households. Some bonuses are applied through the tariff itself and others are applied outside the invoice through social services or help funds.

The first two rates presented, Sevilla and Málaga, are tariffs in which the consumption considered for the definition of the price is *per capita*. The case of the tariff applied in Barcelona differs from the others in its way of determining the fixed fee. The fixed payment that subscribers must make to the water supply service in this city takes into account the type of housing defined by the number of water facilities available. The León rate is included in this analysis, since it is the only one that has been found to have an IRT rate and in which the fixed part depends on a minimum level of consumption.

Another important issue in the comparison of prices paid for water in Spain is the analysis of progressivity. This progressiveness can be measured by taking into account only the variable part of prices (Suárez-Varela *et al.*, 2015), or by considering both the fixed and the variable part. In the latter, a price will be progressive if, when increasing consumption, the average price paid per m^3 is higher. The higher the price being paid per unit depends on how high the level of consumption is on the one hand, and on both the fixed charges being applied and on the structure of the variable part on the other. The larger the proportion of the fixed charges as a total of the bill, the greater the escalation of the variable part to achieve an average price rising in parallel with consumption.

This second approach is the one used in the report by ACA (2015) and is the one adopted in this work. Figure 4.1 shows the percentages representing the fixed and variable parts of the total billing (including all charges associated with the urban

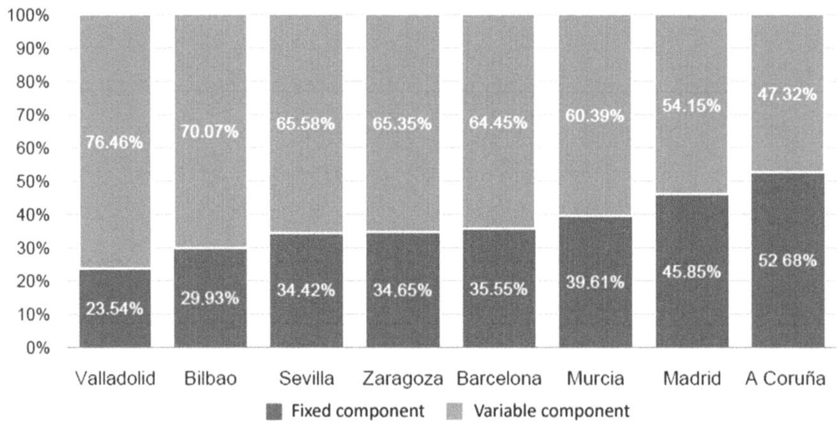

Figure 4.1 Fixed and variable parts of water prices in different Spanish cities (%). Source: ACA (2015).

cycle and VAT) for a consumption of 10 m^3 in eight Spanish cities. The weight of the fixed part goes from approximately a quarter of the bill in the case of Valladolid to more than half of A Coruña. The cities of Seville and Barcelona present a very similar weight of the fixed part despite the significant differences in their price structures (discussed above).

In Table 4.3, the total average price for the consumptions of 6, 10, 12 and 20 m^3 in the eight cities is shown. The cities of A Coruña, Madrid, Bilbao and Murcia present completely regressive prices despite the fact that the percentages of the fixed parts are very different (30% Bilbao and 53% A Coruña). In the case of Barcelona, Zaragoza and Seville, which present very similar percentages of the fixed part as a proportion of the total, prices in Barcelona are much more progressive than in the other two cities.

The diversity in the design of water prices in Spain has a very significant influence both on the progressivity of tariffs and on the price levels paid by water consumers. Various studies have established the average prices paid at the autonomous community, province and demarcation level (AEAS, 2014; García-Rubio *et al.*, 2015) and also at the level of cities and municipalities (ACA, 2015; FACUA, 2016). Certainly, the price paid by a consumer in Spain for a certain amount of water can be very different depending on the municipality in which they live. According to the FACUA report (2016) comparing the average prices of 28 Spanish cities, the differences can reach 350%.

The heterogeneity in water prices derive from the establishment of different legal forms of payments and the multitude of price structures applied. These differences have consequences on the effectiveness of these prices in the fulfilment of their multiple objectives. If we add the existing differences in the billing period, the level of information offered to the consumers and the different competent

Table 4.3 Progressivity of water prices in different Spanish cities.

		Total Price €/m^3 (VAT Not Included)			
City	**CCAA**	**6 m^3**	**10 m^3**	**12 m^3**	**20 m^3**
A Coruña	Galicia	1.77	1.59	1.50	1.36
Barcelona	Cataluña	2.77	2.55	2.65	3.68
Bilbao	País Vasco	1.81	1.54	1.49	1.43
Madrid	Madrid	2.05	1.57	1.45	1.35
Murcia	Murcia	3.78	2.99	2.81	2.43
Sevilla	Andalucía	2.23	2.07	1.96	2.50
Valladolid	Castilla y León	1.13	1.13	1.14	1.23
Zaragoza	Aragón	1.12	1.13	1.13	1.26

Source: ACA (2015).

authorities with diverse modalities of management, we are faced with the need to carry out a local analysis adapted to the reality of each municipality.

4.4 WATER PRICES IN THE CITY OF VALENCIA

The water bill in the city of Valencia includes all the payments that a consumer must make for the services that make up the urban water cycle. This includes payments for supply, sewerage and sanitation. Each of these payments has a different legal structure and nature. Beginning with the phase of drinking water supply both at wholesale and retail levels, the legal nature of the payments made by consumers for these services is that of tariffs or prices subject to private law (Jiménez, 2012). When considering private prices, value added tax (VAT) is applied with a reduced rate because it is water consumption.

The procedure for processing the potable water supply rates consists of two parts since authorisation is required in the higher instance (Álvarez Fernández, 2004). In the first part of the processing, the water management companies request the administration or competent authority (the city council in the case of retail water supply levels and the Metropolitan Hydraulic Services Entity (EMSHI) in the case of wholesale water supply levels) approves a new tariff based on cost recovery, investment and return criteria. This new rate proposal is studied by the municipal services who can request an extension of the information or hold a hearing with the interested parties until they agree to finally approve the rate in the municipal plenary or in the assembly of EMSHI. The second part of the processing falls on the commission of prices of the Valencian community studying the proposed rate; finally, the commission releases a resolution that authorises the approval and publication of the water supply tariffs.

The rates for drinking water in the city of Valencia have a structure in two parts, with a fixed part that depends on the diameter of the meter and a variable part in two blocks with a limit of 12 m^3 bimonthly for the first block. The structure of this tariff is of the IRT type that differs from those usually used in domestic consumption. A differentiation is applied both in the block limit and in the price of each block by the size of the family (large families). Additionally, the water supply fee must be added to the payments for the services of the sanitation phase; this constitutes, on the one hand, a sewer rate that charges a uniform variable price and, on the other hand, a sanitation canon whose structure is also in two parts, the variable part being uniform, that is, equal for all the m^3 consumed. Both the variable part and the fixed part of the canon depend on the size of the municipality that in the case of Valencia corresponds to the section of more than 50,000 inhabitants.

Finally, the Metropolitan Rate for the Waste Treatment and Disposal Service (TAMER) was approved in 2008 and included in the invoice. From its entry into force until its modification in October 2011, the rate was structured in four sections that depended on the water consumption of the household in the previous year (from September to September). The first three tranches consist of a fixed annual quota divided into six bimonthly invoices, while the last section has a structure in two parts with a variable uniform type per m^3 consumed in the previous year (Table 4.4).

Alongside these payments is the investment rate of the city council that is established as a fixed amount per month and the canon the Júcar Hydrographic Confederation established as a uniform variable price. The set of payments that a

Table 4.4 Tariffs, fees and canons applied to domestic water consumption of Valencia (2009–2011).

	2009	2010	2011
WATER SUPPLY TARIFF (RETAIL)			
Service price (€/month)	5.271	5.271	5.271
Consumption price (€/m^3)			
A Basic tariff	0.3094	0.1680	0.1680
B Subsidised Consumption (12 m^3)	0.2321	0.0907	0.0907
C Large families 3–4 children (40 m^3)	0.2166	0.0756	0.0756
D Special large family 5 or more children (120 m^3)	0.0309	0	0
OTHER TARIFFS			
Investment tariff of City Council (€/month)	0.0711	0.0711	0.8300
Water supply tariff (wholesale)			
Costs (€/m^3)	0.178503	0.319881	0.327737
Inversion in facilities (€/m^3)	0.05	0.05	0.05
Hydrographic Demarcation of Jucar (€/m^3)	0.027867	0.027867	0.020011
SEWAGE PRICE			
Sewage canon (€/m^3)			
Fixed charge (€/bimonthly)	5.72	5.72	5.72
Variable charge (€/m^3)	0.342	0.342	0.342
Sewage tax (€/m^3)	0.27796	0.27796	0.282129
TAMER TAX			
To 65 m^3 (€/year)	22.2	22.2	22.2
Between 65–130 m^3 (€/year)	44.4	44.4	44.4
Between 130–195 m^3 (€/year)	66.6	66.6	66.6
More than 195 m^3 (€/year)	88.8	88.8	88.8
More than 260 m^3			
(€/year)	88.8	88.8	88.8
(€/m^3)	0.02	0.02	0.02

Table 4.5 Structure of domestic water prices in Valencia (2009–2011).

	2009	2010	2011
Fixed charges (€/bimester)	16.4042	16.4042	17.922
Variable $< 12\ m^3 (€/m^3)$	1.10843	1.108408	1.112577
Variable $> 12\ m^3 (€/m^3)$	1.18573	1.185708	1.189877

domestic subscriber makes in the city of Valencia is summarised in Table 4.4 for the period 2009–2011. To calculate the fixed fee (service fee), one must depend on the meter diameter; the most common one in the city of Valencia is 15 mm, and represents 95% of the total number of meters installed in Valencian homes.

This set of payments made by the consumer in each invoice can finally be summarised in a simple structure. The final structure of the water price consists of a fixed part and a volumetric part ($€/m^3$) that will be different depending on the subscriber consumes in the bimester. To this price structure (Table 4.5) we must add the effect of the TAMER, thus adding a fixed charge per invoice depending on the previous year's consumption or a fixed charge plus another variable if the subscriber's consumption exceeded 260 m^3.

As a result of the application of this structure, average prices are always higher than marginal prices for to two reasons. On the one hand, the fixed part is very large with respect to the variable part and, on the other hand, the variable part is not particularly progressive. The relationship between the fixed part and the variable is represented in Figure 4.2, in which the percentage that represents the fixed part is around 60%. On the other hand, this percentage has increased between 2009 and 2011.

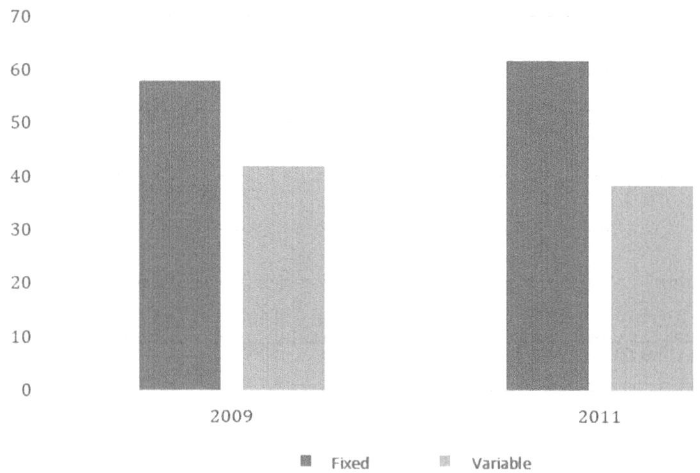

Figure 4.2 Fixed and variable parts of water prices in Valencia (2009–2011) (%).

Table 4.6 Progressivity of water prices in Valencia (2009–2011).

Water Consumption (m^3)	6	10	12	20	50	100
2009	3.84	2.75	2.48	2.01	1.51	1.35
2010	3.84	2.75	2.48	2.01	1.51	1.35
2011	4.10	2.90	2.61	2.09	1.55	1.37

In order to measure the progression of the structure applied in Valencia, a simulation of the average prices for the consumptions of 6, 10, 12, 20, 50 and 100 m^3 has been made to the bimester applying all the tariffs and charges that appear in bimonthly bills. Neither the VAT nor the charge corresponding to the TAMER is included. As can be observed in Table 4.6, lower consumption levels pay a higher average price than higher consumption levels. In this sense, the tariff applied in Valencia is regressive since, for all consumption levels considered, the average price is higher than the marginal price.

The lack of progressivity in the tariff is due to the application of an IRT and not an IBT, to the small price difference that exists between the two consumption blocks, and to the existence of only two blocks with a very small limit (12 m^3), especially if we take into account that this limit is for housing and not per capita.

4.5 CONCLUSIONS

The Water Framework Directive (WFD) establishes the need to use prices as an instrument to recover costs and as an incentive to improve efficiency contributing also to the achievement of the environmental objective. Given that the supply of drinking water is a municipal responsibility, in Spain there exists a great heterogeneity of price structures. As an example of the relevance of these differences in payments for water the tariffs of some Spanish cities are analysed.

One important issue in the comparison of prices paid for water in different areas is the analysis of progressivity. This progressiveness can be measured by taking into account only the variable part of prices or by considering both the fixed and the variable part. In the latter, a price will be progressive if, when increasing consumption, the average price paid per cubic meter is higher. The diversity in water prices is caused by the different forms of payments and the great variety of price structures applied. In addition, it should be taken into account the existing differences in the billing period, the level of information offered to the consumers and the different water authorities with diverse management criteria. Because of that, the study of water prices requires to be carried out at the local level.

In the case of the city of Valencia, the price structure has two parts: one fixed part that depends on the diameter of the meter and a variable part in two blocks with a limit of 12 m^3 bimonthly for the first block. It is a water-pricing structure known as Increasing Rate Tariff (IRT), that is, the payment depends on the consumption, but a higher price per unit is paid for the higher the consumption level and all the units are paid at the same price. In addition, the payment for sanitation services must be added to the water supply rate. Its structure is constituted by a sewer rate that charges a uniform variable price and a sanitation canon whose structure is also in two parts, the variable part being uniform. In order to measure the progressivity of the price structure applied in the city of Valencia, a simulation of the average prices for different levels of water consumption has been made applying all the tariffs and charges that appear in bimonthly bills.

REFERENCES

ACA. (2015). *El precio del ciclo del agua en España y Europa 2015* (The price of the water cycle in Spain and Europe). ACA, Barcelona. (In Spanish).

AEAS. (2014). *Tarifas 2012. Precio de los servicios de abastecimiento y saneamiento en España* (Tariffs 2012. Price of water supply and sanitation services in Spain). AEAS, Madrid. (In Spanish).

Álvarez Fernández M. (2004). *El abastecimiento de agua en España* (The water supply in Spain). Civitas, Madrid. (In Spanish).

Arbués F. and Barberán R. (2012). Tariffs for urban water services in Spain: household size and equity. *International Journal of Water Resources Development,* **28**(1), 123–140.

Collins R., Kristensen P. and Thyssen N. (2009). Water Resources Across Europe-Confronting Water Scarcity and Drought. EEA Report No. 2/2009. Official Publication of the EU, Luxemburg.

EA. (2000). Comunicación de la Comisión al Parlamento Europeo, al Consejo, al Comité Económico y Social y al Comité de las Regiones. "Política de tarificación y uso sostenible de los recursos hídricos". 477 final, European Commission, Brussels, 26 July, 2000. (In Spanish).

EEA. (2010). The European Environment. Stater and Outlook 2010. Adapting to Climate Change. European Environment Agency, Copenhagen, Denmark.

EEA. (2012). Urban Adaptation to Climate Change in Europe. European Environment Agency, Official Publication of the EU, Luxemburg.

FACUA. (2016). *Estudio comparativo sobre las tarifas de agua en 28 ciudades* (Comparative study on water tariffs in 28 cities). FACUA, http://www.facua.org/es/guias/estudio-tarifas-agua2013.pdf (In Spanish).

Fuentes A. (2011). Policies Towards a Sustainable Use of Water in Spain. OECD Working Papers No 840. OECD, Paris.

García-Rubio M. A., Ruiz-Villaverde A. and González-Gómez F. (2015). Urban water tariffs in Spain: what needs to be done?. *Water*, **7**(4), 1456–1479.

García-Valiñas M. Á., Martínez-Espiñeira R. and To H. (2015). The use of non-pricing instruments to manage residential water demand: What have we learned? In: Understanding and Managing Urban Water in Transition, Q. Grafton, K.A. Daniell, C. Nauges, J. Rinaudo and N. Chan (eds), Springer, Berlin, pp. 269–281.

Jiménez I. (2012). *El régimen económico financiero de la distribución de agua* (The economicfinancial regime of water distribution). In: *Actas Jornadas de Derecho de Aguas Zaragoza, Agua y Ciudades, Civitas* (Proceedings of Water Law

Conference. Water and Cities. Civitas. Zaragoza), A. Embid- Irujo (ed.), pp. 79–162. (In Spanish).

Komives K., Foster V., Halpern J. and Wodon Q. (2005). Water, Electricity, and the Poor: Who Benefits from Utility Subsidies? World Bank Publications, Washington DC.

Labandeira X., León C. J. and Vázquez M. X. (2007). Economía Ambiental. Pearson Education, Madrid. (In Spanish).

Massarutto A. (2007). Water pricing and full cost recovery of water services: Economic incentive or instrument of public finance? *Water Policy*, **9**(6), 591–613.

OECD. (2003). Social Issues in the Provision and Pricing of Water Services. OECD, Paris.

OECD. (2009). Managing Water for all. An OECD Perspective on Pricing and Financing. OECD, Paris.

OECD. (2010). Pricing Water Resources and Water and Sanitation Services. OECD, Paris.

Sartorio S. (2009). "*La tributación del Agua*" (Taxation of water). In: *El Agua: Estudios Interdisciplinares* (Water: Interdisciplinary studies). M.J. Montoro (ed.), Institut de l'Aigua, Barcelona, pp. 231–241. (In Spanish).

Strosser P., Delacámara G., Gómez C.M., Lago M. and Maziotis A. (2013). Changing current practice in the application of EPIs to achieve the objectives of the WFD; issues and options. EPI-WATER – Evaluating Economic Policy Instruments for Sustainable Water Management in Europe. EPI-WATER Discussion Paper 01, September 2013.

Suárez-Varela M., Martínez-Espiñeira R. and González-Gómez F. (2015). An analysis of the price escalation of non-linear water tariffs for domestic uses in Spain. *Utilities Policy,* **34**, 82–93.

Zetland D., Dworak T., Lago M., Maziotis A. and Viavattene C. (2013). An orientation to the assessment framework for evaluating economic policy instruments for managing water. EPI-WATER. Evaluating Economic Policy Instruments for Sustainable Water Management in Europe. Discussion Paper 02. September 2013.

Chapter 5

Tariffs and cost recovery

*Mónica Pueyo, Oriol Figueras and Mariona Coch**

Tax and Revenue Department, Catalan Water Agency, Provença, 260, 08008 Barcelona, Spain

**Corresponding author Email: mcoch@gencat.cat*

ABSTRACT

Water prices in Catalonia have grown noticeably since the principles and objectives of the Water Framework Directive were set. The Catalan Water Tax, an environmental tax created in the year 2000, has fostered the use of progressive structures in water tariffs, set by the municipalities. These structures encourage the efficient use of the resource. However, full cost recovery has not been generalized yet in all municipalities, because maintenance and renovation cost of infrastructures are underestimated.

5.1 INTRODUCTION

The Water Framework Directive (2000/60/EC) (European Community, 2000) which applies to all EU member states, stipulates that economic aspects play a key role in protecting and improving water and aquatic ecosystems and outlines two arguments. First, the directive stresses the importance of carrying out an economic analysis of water use as part of the planning process, since this makes it possible to study the various alternatives and ensure that decisions concerning the actions and measures to be carried out are based on economic rationale rather than timeliness. Second, the directive also suggests that a sustainable, balanced

Editor(s): Francesc Hernández-Sancho
doi: 10.2166/9781780404684_0071

and equitable water supply that meets the needs of people and economic activities can only be guaranteed by proper cost recovery through prices. Member states are required to establish price mechanisms as a means of achieving targets relating to the reduction of water pollution, the preservation of water bodies and protected areas, and other environmental measures.

Countries have attempted to overcome this challenge in different ways, depending on their social and economic realities, the actors involved in the water cycle and the organizational structures adopted for water management. This means that there is no single approach to the subject, just as there is no miraculous solution that can be applied to every location and situation. It is necessary to put a fair price on water that meets social and territorial objectives that does not simply internalize all service costs, but sends signals to the market about the environmental costs required to achieve good water status. In addition, the costs of the resource must be taken into account so that future generations have access to water and can enjoy an environment in as good a state as possible. The pursuit of this ambitious objective relies on a number of instruments, including tariffs, charges and taxes for sanitation services, water supply, sewage services and reused water, which together make up the price of water. In order to carry out the process of setting the end price, three factors that have determined the actions carried out over the last 25 years must be taken into account.

5.1.1 Water service costs

Water service costs have increased significantly in recent times. This is due to the scarcity of the resource, the growing concern for quality (as evidenced by stricter regulations), greater environmental constraints and, lastly, the devastating effects of climate change, which is causing recurrent droughts and floods with much greater severity than in the past. All of these factors result in a greater need for costly measures and treatment processes, and advanced technologies for transportation, water treatment, desalination and reuse.

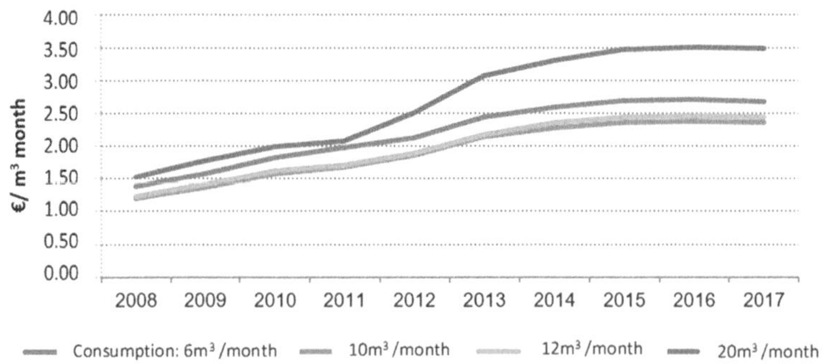

Figure 5.1 Evolution of the water price of the water cycle (weighted average per inhabitant).

Despite significant increases in water prices (Figure 5.1), the values demanded by service operators have not been achieved, at least in most Mediterranean countries. Although water is scarce in the region, this was traditionally not reflected in the end prices, which were greatly reduced and far removed from the actual costs. Thus, consumers in this region have traditionally benefited from low prices, and many of the water companies have been unable to offer anything beyond defective, subsidized and inefficient services, because they have fallen into the following vicious cycle: low price ⇒insufficient financial resources to provide the service ⇒ lack of investment ⇒ decline in service ⇒ low price.

5.1.2 Waste-water treatment

Waste-water treatment was not regarded as a problem by society, beyond the health aspect, until the end of the 20th century. Even today, compliance with European regulations throughout the European Union cannot be fully substantiated. Few local authorities take control of sewers or networks to collect municipal waste water before it is discharged into treatment systems, if they are not directly responsible for such systems, so this activity takes a back seat in favor of other municipal responsibilities. As a result, the resources allocated by municipalities to investment and exploitation in this phase of the cycle are minimal, and the cost of inaction is transferred to the next phase.

This lack of action drives up the cost of water treatment systems because the discharge of pollution is not prevented at source. As a consequence, the cost of treatment is borne by all users, rather than by those responsible for the most polluting discharges. It is also necessary to consider the high cost of treating this sludge, which often contains high levels of chemicals, due to the lack of control over discharges, and therefore cannot be used as compost or for any other form of recovery. This highlights the fact that the polluter-pays principle, which is widely accepted by society, requires the costs to be properly shared among the various actors and also calls for greater managerial efficiency, not only by promoting the construction of infrastructure and improving its performance, but by focusing on the root of the problems and carrying out genuine preventive action.

5.1.3 Environmental protection

Environmental protection, as a key element to ensure that today's population and future generations (with due consideration for their inherent economic activity) have complete and sustainable access to water, is now accepted even by traditionally more reluctant sectors, which have realized that it is not merely an "environmental" measure, but also, and above all, a measure that can be cost effective in the long term. Passing on the cost of environmental actions to citizens through fiscal measures is a widespread practice in modern countries, either through taxes, charges or other levy types, and aims to modulate the behavior of citizens and companies and foster eco-efficiency through incentives. There are also major topics for discussion in this area, especially with respect to the nature and characteristics of these fiscal instruments and how they are used to achieve the objectives pursued. If taxpayers understand the rationale behind these economic measures, they are more likely to accept the water management process.

5.2 FACTORS DETERMINING WATER SERVICE PRICES

Developed countries incorporate water costs into prices to varying degrees, with due consideration for the above factors, and use a number of different organizational, management and regulatory instruments, forms and systems to tackle this challenge. A non-exhaustive list of the main items that require analysis in order to correctly implement a pricing policy and comply with the Water Framework Directive is given below.

The costs arising from the operation and maintenance of water services must be incorporated into tariff proposals to ensure that consumers are ultimately responsible for paying them in full through the water bill. There is generally no room for maneuver in the amounts paid for goods or services rendered by third parties if they are the result of public procurement procedures or processes in which publicity and competition have been guaranteed. Costs related to companies' internal data, based on estimates and pro-rata calculations provided by the water supply companies themselves, require a certain degree of formalization to ensure the suitability of the different allocations.

Sometimes "official" reference prices or values can provide criteria for the reliability of the amounts involved, but many items present a wide margin for discretion that must be assessed by supervisory or regulatory bodies, depending on whether the service provider is public or private. The current trend is to share information between the different services to identify the level of costs in the sector and analyze the differences between them. Thus, the use of indicators can provide warning signals that should give rise to a detailed analysis of the factors that influence each service.

This is the case for electricity costs, which are usually related to market prices. However, regulation of the Spanish electricity sector allows for plenty of room for negotiation to reduce prices, so they can vary. When the costs of different operations with similar services are compared, it becomes clear whether or not there is room for improvement. In addition to administrative and legal improvements aimed at enhancing competition in procurement, a number of technical measures are also essential to cut electric costs. This often involves work to optimize and adjust processes and facilities that is essential for revising the energy model, and this requires a level of knowledge and specialization that is beyond the capacity of many operators. Public and private companies and associations have proliferated that provide advice and guidance, and jointly bid to obtain more competitive energy prices.

On top of this, personnel costs involved in water-cycle services should be high enough to guarantee an efficient management process. When the service is provided by the public sector, these costs are often underestimated. This occurs because the costs required to carry out the activity are not taken into account, since it is partly carried out by municipal staff workers responsible for many different tasks who may perform some water service functions from time to time. In some cases, the opposite may occur and overstaffed services or excessively high salaries per employee have been found.

It is difficult to assign the cost required to provide a good service without having an in-depth understanding of its characteristics, but the average amount per employee of each company makes it possible to compare it to the pay level of the sector. Once the average cost per employee has been analyzed, a second question arises as to whether the legal measures for deficit reduction implemented by the Spanish government in 2010, which gave rise to financial regulation of the public sector, are applicable to maximum salary levels. These measures stipulated the upper salary limits in this sector. When applying these measures, the Catalan Parliament echoed the Spanish provisions and, through Additional Provision 21 of Act 2/2014 (Government of Catalonia, 2014a), established that the full annual remuneration of public sector managers must not exceed the limits generally established for the position of minister of the Catalan government and should, as a general rule, be in keeping with the remuneration levels established for senior positions in the Catalan government. Because current legislation stipulates an upper salary limit for workers in the public sector, it seems illogical that sums exceeding these amounts should be charged to consumers through tariffs. In no way does this imply that private concessionaires, which are not subject to these provisions, cannot remunerate their management personnel in line with market prices, if these are higher than those provided for in the regulations, but the difference between the statutory maximum and the salary paid should be paid for through company profit.

In Catalonia, where large regional networks provide water to supply companies so that they distribute water to a significant percentage of the population, the amounts paid to the public or private wholesale operators of these regional networks for the purchase of water represent an important part of the cost structure of some water supply companies

This subject has been systematically addressed through the regulatory process the Catalan Water Agency is required to implement in accordance with current legislation, which covers, in addition to technical water management, the approval of wholesale tariffs (retail tariffs are authorized by the Catalan Pricing Commission (Government of Catalonia, 2014a). This is due to the unique strategic characteristics of the service provided by large regional networks, which call for special treatment that allows the water authority not only to intervene in determining prices, which are passed on to end users, but also to manage and prioritize the measures set out at the planning stage and guide the development policy carried out. The aim of this regulatory process is to influence the behavior of the different stakeholders, promote actions and create more balance in the territory, thereby fulfilling the main goal, which is to guarantee the water supply for the population.

Outsourcing or work carried out by third parties is increasingly being included in tariff proposals. Such work is required to meet ad-hoc technological and specialization needs that are difficult to handle using the company's own resources. To keep such costs in check, the criteria and principles governing procurement must be respected so that these services are allocated at competitive or market prices. Excessive outsourcing in strategic areas or areas inherent to the company's activity is only justified if it is proven to help reduce costs or optimize management.

Investment in infrastructure and machinery that experts say must be carried out over the next few years in the sector will be even greater partly due to abandoned action plans and delayed implementation, but also to the need to comply with a whole host of new requirements on environmental, health, safety at work, technological, social needs, etc. Thus, in addition to traditional needs such as greater investment and upgrading of facilities, property and equipment, new measures are required to comply with current legislation that is revised on a constant basis, as well as measures associated with the technological changes that help achieve excellence in water management.

One of the main features that characterizes the water sector at the moment is the high demand for investment, since the measures required call for huge capital injections and a long-term return on investment. Such major infrastructure projects and expensive equipment require financial contributions that are beyond the means of many municipalities and local authorities, which generally have insufficient resources and have exceeded their external financing quota. For that reason, this issue is often resolved through the adoption of public-private partnerships in which private partners can help access financial markets. However, the result is variable due to the asymmetry between the interests, principles and aims of partners.

It is also necessary to be aware of the requirements of the anti-corruption, economic transparency and public procedure policies adopted by many countries, especially in a sector such as this, which involves the movement of large sums of money. An example of legislation aimed at combating certain practices is the Sapin Law of 1993, a second version of which was promulgated by the president of France on December 9, 2016 (French Constitutional Court, 2016). This law has had a positive impact by improving the control and supervision systems of public services and shortening public procurement periods to promote competition and, ultimately, to ensure that citizens receive better-quality services.

Although the relationships between local authorities and their concessionaires are regulated by contractual conditions, only by understanding these conditions down to the smallest detail is it possible to make assertions about certain aspects of the service, especially the price. However, some ideas should undergo critical scrutiny, despite the fact that they have been accepted in the past. First, it is generally believed that, unless the contract expressly states otherwise, infrastructure resulting from investments undertaken by service operators belongs to them, even when the annual depreciation expense has been included in tariffs, i.e., users have paid for it.

Although it is true that the infrastructure owner is responsible for depreciation, the inclusion of this cost in tariff proposals, and therefore in the price of the service, grants consumers certain rights over the asset. In the event that the water services concession period ends and it is decided that the concessionaire must be paid for infrastructure, consumers end up paying twice for the same asset.

Another important issue lies in the treatment of the depreciation of infrastructure belonging to local authorities. Many tend not to set aside enough funds to recapitalize the service and pursue self-financing when the facilities come to the end of their useful life. It is therefore argued that, since local authorities do not have the funds to renew infrastructure and have very limited access to capital markets, they are required to approach regional, national or European authorities to request subsidies to help them finance new investments.

In this context, infrastructure depreciation of public and private companies should be included in tariff proposals, so as not to distort prices or create grievances between entities that do not receive subsidies from the authorities and those that do. At the same time, by applying current accounting procedures, subsidies should be included in tariff proposals so that a neutralizing effect can be achieved with respect to the subsidized part by offsetting expenditure (investment) against income (subsidy).

It is of course unreasonable for local authorities to play the victim and benefit from public aid while keeping prices artificially low and not taking measures to improve the situation and their self-financing capacity. It is also true that the effect of economies of scale must be taken into account, in the sense that some infrastructure has so few users that transferring the investment cost to the price of the service (the non-subsidized part only) would result in unacceptable costs for users. This fact brings us back to the need to find optimal units for providing service to reduce investment and operational and maintenance costs, and ensure that this reduction is reflected in the final price.

There can be no denying that each subsidy process is designed to alleviate the most poorly resourced economies and that granting aid implies the tacit acknowledgment that the recipients are not able to afford the required actions themselves. This means that public water aid criteria must take pricing policy into account to ensure that efficient use of the resource is encouraged and to contribute to making the principles of the Water Framework Directive effective.

This reasoning applies to costs of water works and machinery, and also to studies, projects and facilities that have been financed by public subsidies.

Tariff proposals frequently include the fee paid by the concessionaire. This is the amount paid by the operator to the owner of the facilities used for the public service as payment for using them. In most cases it consists of a small percentage of the fixed tariff or an amount based on the volume supplied, but sometimes it requires the advance payment of enormous sums that can hinder competition. Justification of this payment is related to the capitalization of depreciation for use of the infrastructure, but the reality is that cash injections involving huge amounts of money prevent some candidates from submitting tenders, thereby suppressing the beneficial effects of competition in bidding processes.

The last item to be addressed is the remuneration for the operator or concessionaire, which is generally included in tariffs, to ensure that the operator or concessionaire receives the amounts legally due to them for fulfilling their functions. Remuneration for the contractor, partner or operator is normally set out in the agreements, contracts or documentation that regulate the relationship between the service owner and provider. It usually takes the form of a percentage of the operating expenses, so that the higher the expenditure, the higher the payment. This can have an adverse effect on the approach, at least on paper.

Recent trends include the application of objective criteria, determined based on the volumes treated or supplied, with some scope for including aspects that reflect the efficiency and continuous improvement of the service. The amount of this remuneration is another topic for debate. For many years, the range of percentages has varied widely (3–16% of Operational Expenditures, OPEX). High rates are perhaps appropriate for industrial activities involving major risks, but may not be suitable for such a monopolistic sector, and, although the agreements signed by the parties must, in principle, be respected, profits should be subject to fairly restrictive interpretation.

In summary, although this is not an exhaustive list, it includes the most contentious aspects of the water pricing process. Most of the criteria are applicable to water supply tariffs and should even be taken into account when establishing taxes that occasionally serve to offset services such as water sanitation and waste-water treatment and reuse. Within this general analysis of the pricing policy, it should be noted that the scale of the service unit is key to determining the final price of the service. From the moment that authorities seek to guarantee that the population has access to water and to carry out the waste-water treatment process that is inherent to the service, there is a need for optimal action areas to ensure that the quality of the service and the costs and prices it entails are sound and tend to be homogeneous. Recent Spanish legislation stipulated the need to optimize service provision areas, but the solution to this problem has not been addressed with the same success as seen in other countries, where effective policies and instruments are used to promote the grouping of public services into areas that favor the service's sustainability and technical and economic efficiency.

5.3 WATER SERVICES MANAGEMENT IN CATALONIA

In order to guarantee supply with the highest levels of quality, efficiency and sustainability, the entire water cycle must be considered a single unit: water catchment and purification, distribution for consumption, and, finally collection and treatment of waste water so that it can be returned to aquatic ecosystems without harming the environment.

5.3.1 Scope of competence

Responsibility for the complete water cycle in Catalonia is shared among numerous public and private actors.

5.3.1.1 Local authorities

In Spain, municipalities have competences in the area of water supply and sanitation, and this is reflected in regional legislation, specifically Article 66.3.I of Legislative Decree 2/2003, of April 28 (Government of Catalonia, 2003a), approving the consolidated text of the municipal and local regimen of Catalonia. Furthermore, Article 5 of Legislative Decree 3/2003, of November 4 (Government of Catalonia, 2003b), approving the consolidated text of the legislation relating to water in Catalonia (TRLAC), outlines the scope of competence of local authorities in water-related matters covering: (a) the supply of drinking water; (b) sewage and waste-water treatment; (c) sanitary control of waste water; and (d) performance of the duties conferred upon them by this law.

The local authority exercises its competences to provide the service and therefore approves the pricing or tariff system, including the tariff structure. Local public services may be managed directly or indirectly. Direct management can take the following forms: (a) management by the local authority itself; (b) an autonomous body; (c) a trading company with public capital. Indirect management can take different forms: concessions, cooperative arrangements between the public and private sectors, and different forms of outsourcing public services, including cooperatives with mixed private-public capital.

5.3.1.2 Catalan water agency

The Catalan Water Agency (ACA) is a public company that operates under the Ministry of Planning and Sustainability of the Catalan government and is responsible for managing the complete water cycle, in accordance with Legislative Decree 3/2003, of November 4 (TRLAC). The holistic approach of the Catalan Water Agency focuses on aquatic ecosystems and strives to ensure they are in equilibrium, and is based on the principles of efficacy, efficiency and cost effectiveness. It implements the Catalan government's policy on water, and its primary objective, in accordance with work plans, is to ensure the proper management of the water cycle and to guarantee the availability and quality of water at its source and the preservation of water ecosystems. It also plans, organizes and finances urban waste-water sanitation throughout Catalonia.

With respect to Catalonia's internal basins (not including the Ebro and Jucar river basins), it also serves as a river authority and is therefore responsible for the regulation of reservoirs, and the management and concession of water resources and their exploitation in the public domain. This organization is financed primarily through an environmental tax, i.e., the water tax, which is charged by ACA or included in the bill when water is furnished by a supply company. Proceeds from the tax are used to finance the construction and operation of water cycle facilities, hydraulic infrastructure and measures to control, protect and preserve water ecosystems.

5.3.1.3 Pricing Commission

The Pricing Commission is the body that authorizes maximum prices, at the regional level, in the area of water supply services, urban passenger transport and railway companies, as part of the powers conferred on it by Decree 121/2014, of August 26, of the Catalan Pricing Commission (Government of Catalonia, 2014c). This provision repeals Decree 164/1993 of June 1, as amended by Decree 251/1996 of July 5. The Pricing Commission reports to the General Secretariat of the Ministry of Territory and Sustainability.

Water tariffs are approved by the relevant local authority and the tariff proposal is then submitted to this body so it can carry out its pricing policy functions. The proposal includes the agreement of the local council, as well as an economic analysis and any supporting documentation. Under the simplified regime procedure, tariffs may be increased by a percentage that is the same as or lower than then one approved annually by the Pricing Commission with no need for prior authorization.

5.3.1.4 Water price components

The water bill includes the supply fare, the water tax (the tax of the Catalan government) and VAT, as well as other items such as sewage. In some municipalities, the water bill also includes items that are not directly related to the water cycle, such as waste management fees. This revenue is not included in the following calculations of the total water price. Figure 5.2 shows the percentage of each component in the water bill (for a consumption level of 12 m^3 per subscriber and month); the water supply tariff represents 56% of the bill, followed by the water tax (26%), sewage services (8.4%) and then VAT.

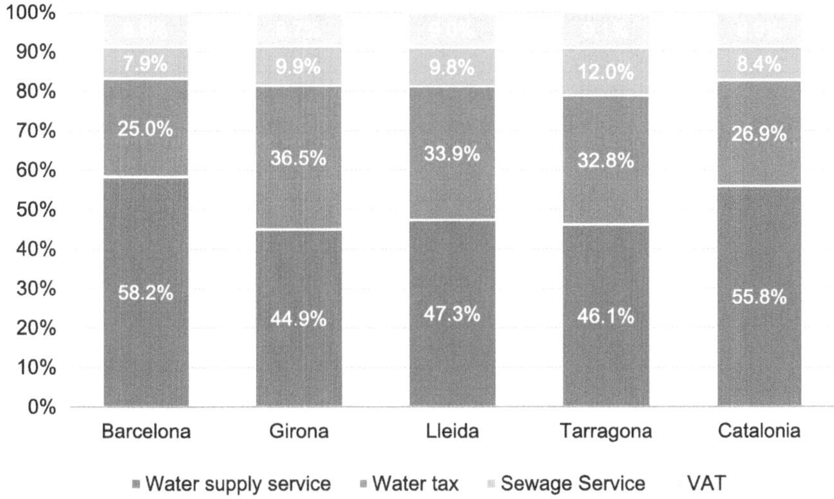

Figure 5.2 Percentage of water price concepts depending on the province (VAT included).

5.3.1.5 Domestic water tariff

Domestic water supply tariffs must be established so that the service is self-financing, as set out by the European Water Framework Directive (water catchment, transport and treatment, among others). It is therefore essential to ensure that the service is economically and financially balanced. On rare occasions, circumstances may result in tariffs that do not cover the cost of services, but this must be justified.

The operator can design a tariff structure that passes on the service costs according to consumer type. This may be based on housing type and water use, and may include social measures to guarantee access to the resource for all. The basic structure of a domestic water tariff usually includes a fixed component, or fixed charge, and a variable component. It is also possible to include a minimum consumption level or billable amount. The minimum billable amount involves applying a certain volume, even when that volume has not been consumed. Once this volume is exceeded, users are billed for their actual consumption.

Including a fixed component in the water bill is designed to guarantee a minimum level of funds to cover the expenses incurred, regardless of consumption level, since many of the expenses of the water service are fixed and non-variable. Some entities apply a minimum billable amount, others apply a charge, and some do both. Minimum billable amounts and charges are usually linked to a guaranteed service for citizens. It is important to note that, because of the fixed components of the tariff, i.e., the minimum billable amount and charge, the bill does not go down as much as one would expect when consumption is reduced. Aside from the supply

Table 5.1 Catalan Municipalities and inhabitants charged with a minimum billable amount or fixed charge.

Tariff Structure	Municipalities with > 5,000 Inhabitants (%)	Inhabitants of mun. > 5,000 Inhabitants (%)	Municipalities Catalonia (%)	Inhabitants Catalonia (%)
Minimum consumption level	19.16	7.44	34.32	10.28
Fixed charge	76.64	91.19	48.26	87.12
Minimum and fixed charge	3.27	1.18	4.54	1.52
No minimum or fixed charge	0.93	0.19	12.88	1.08

Source: ACA (2017).

tariff, other items include meter and connection maintenance, and surcharges for funding hydraulic works and facilities, etc. These items are also included in the calculations to determine the price of the water supply service in Catalonia. Table 5.1 shows the percentage of municipalities and inhabitants charged either a minimum billable amount or a fixed charge, both components or neither component.

With respect to the variable component, which is proportionate to the volume of water supplied, most prices are determined per cubic meter and differ depending on the consumption level, so that the more water consumed, the higher the price per cubic meter (Table 5.2). Only 1% of municipalities with more than 5,000 inhabitants do not take into account consumption blocks or only include one variable block in their tariffs. Most municipalities apply more than three blocks in the variable portion.

5.3.1.6 Water tax

The objectives of the water tax are to encourage more rational and efficient water use, to promote technological innovation, to improve industrial processes and economic activity, and to ensure financial sufficiency and managerial autonomy. The water tax can only be used to finance activities and actions aimed at: pollution prevention at source, recovery of aquatic ecosystems and maintenance of ecological flows; achievement of hydrological planning objectives, particularly the funding of investment and operating costs associated with planned infrastructure; or the authority's own management and financing costs. The revenue raised by the water tax in 2017 was €462 million, 63% of which corresponded to domestic use and 37% to industrial uses and other economic

Table 5.2 Percentage of municipalities and citizens according tariff structure blocks.

Tariff Structure	Municipalities with > 5,000 Inhabitants (%)	Inhabitants of mun. > 5,000 Inhabitants (%)	Municipalities Catalonia (%)	Inhabitants Catalonia (%)
Zero consumption blocks	0.47	0.08	6.96	0.48
One consumption block	0.47	0.14	7.17	0.47
Two consumption blocks	9.35	4.01	17.30	5.04
Three consumption blocks	37.85	25.04	40.82	27.47
Four consumption blocks	33.64	22.70	20.05	22.47
More than four consumption blocks	18.22	48.03	7.70	44.07

Source: ACA (2017).

activities; the ACA's other revenue (from other taxes and fees) amounted to €12 million (Figure 5.3).

The water tax rate is the same throughout Catalonia, with one rate for domestic users and another for industrial users operating under the ordinary regime (where the water tax is not calculated based on the degree of pollution they discharge). There is also a special regime for industrial users based on the amount of pollution discharged. However, a demographic concentration reduction coefficient is applied to domestic consumers from municipalities with no ACA-financed waste-water treatment.

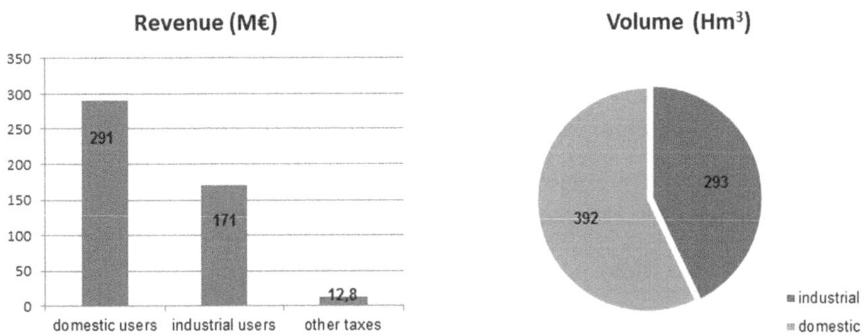

Figure 5.3 Catalan Water Agency revenues and corresponding consumption volume. Source: ACA (2017).

5.3.1.7 Tax rate for domestic uses

The water tax for domestic use is defined for residential, private or communal uses, by individuals or even bodies corporate, corresponding to water used for toilets, showers, cooking and dining, washing clothing and crockery, irrigation of gardens, swimming pools and other communal areas, refrigeration and air conditioning, and other uses that can be considered inherent or specific to human activity in homes. The water consumed in the domestic setting has one of two sources: the water supply network – a water tax is applied and passed on to subscribers by the water supply companies (municipal or private) in the water bills they issue; or the user's own source (wells and other facilities) – the water tax is paid by the Catalan Water Agency and the owner of the water source must carry out the required procedures with this authority (initial declaration of details and quarterly declaration of readings).

The water tax is calculated according to the volume of water consumed, preferably determined by individual meters. The tax rate is affected by increasing coefficients as the blocks are exceeded, so that, the more water consumed (m^3), the higher the tax rate. There is a minimum charge of 6 m^3 per subscriber and month. The tax structure is designed to promote the efficient use of water and encourage consumers to reduce their consumption, given that the system benefits those who consume less. Consumers are subject to the protected rate (€0.4936/m^3), provided that their consumption does not exceed the 1st block (Table 5.3). For those whose consumption level falls within the 2nd, 3rd or 4th block, the general rate is applied (€0.5685/m^3 × consumption coefficient).

Figure 5.4 shows the distribution of consumption blocks by volume and amount billed for 2017. The average value of the domestic water tax was €0.764/m^3, based on declared consumption by companies.

5.3.1.8 Extension of water tax blocks

As mentioned above, the water tax is a tax on water consumption and is calculated based on blocks. It is subject to progressive rates, so that the more users consume, the more they pay. The first consumption block is based on the basic water supply set out in the first additional provision of Legislative Decree 3/2003 (Government of Catalonia, 2003a). To ensure that households with a higher number of members than the average in Catalonia are treated fairly, provisions have been made to extend

Table 5.3 Increasing blocks rates.

	1st Block	2nd Block	3rd Block	4th Block
Volume (m^3)	$<0=9$	10–15	16–18	>18
Price (€/m^3)	0.4936	0.5685 × 2 = 1.137	0.5685 × 5 = 2.8425	0.5685 × 8 = 4.548

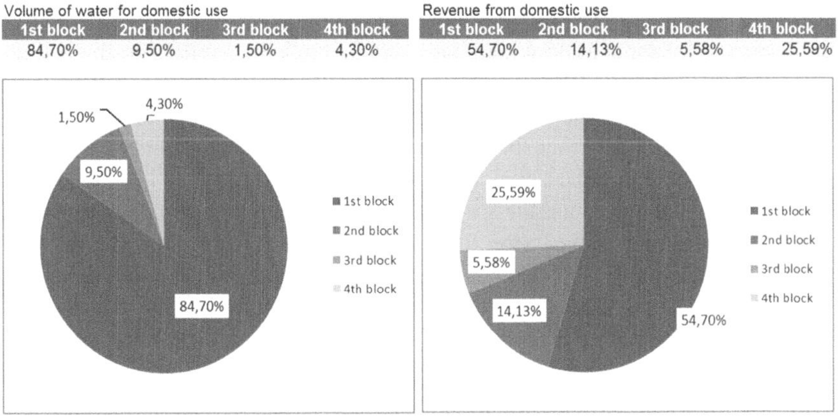

Figure 5.4 Distribution of water consumption by volume and billable amount. Source: ACA (2017).

the water tax blocks to allow an extra 3 m^3 per person per month to be added to the water consumption limits for the 1st block, 2 m^3 per person per month for the 2nd block and 1 m^3 per person per month for the 3rd block. This discount applies to households with four or more members, and special treatment is also available for households with individuals who have a degree of disability greater than 75%. For the purposes of the extension, these individuals count as two people. When these individuals join households with four or more members, an extension can also be requested.

The following graph (Figure 5.5) shows the effect of the blocks and their extension in the case of households with four and five people.

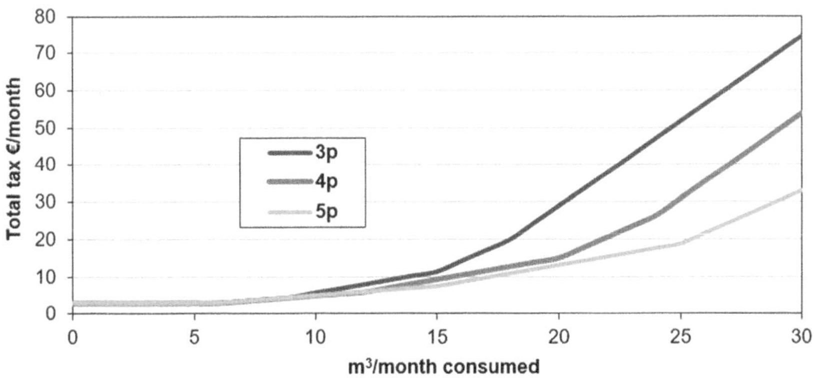

Figure 5.5 Monthly tax for domestic use for 2017.

5.3.1.9 Tax rate for industrial and similar uses

The water tax for industrial and similar users is applied according to the activity carried out, the volume of water used and the pollution generated in the course of this activity. The water tax applied to industrial and similar activities is determined based on whether the activities are subject to the ordinary or special tax regime. In both regimes, the rate is the sum of the general tax rate, which corresponds to water use, and the specific tax rate, which corresponds to pollution. The specific tax rate is the same for all establishments that pay tax under the ordinary regime, while establishments that pay tax under the special regime are subject to an individualized tax based on the amount of pollution they discharge.

The general tax rate for 2017 is €0.1654/m^3. This rate is affected by a number of coefficients, depending on the use of the water: agriculture, livestock, water supply, energy production, aquaculture, etc. The specific tax rate for 2017 for activities under the ordinary income tax regime is €0.6486/m^3, while it is determined individually for activities under the special regime, depending on the amount of pollution discharged, in accordance with the values of the following pollution parameters (Table 5.4).

The formula for calculating the specific tax rate for the special regime is as follows:

$$Te = [\Sigma n(\Sigma y((Ci - Ei) \times Pui \times Cpi \times Fsi \times Kdi \times Kni) \times Kan \times Fn \times Rpn)]$$

where Te = the specific tax rate applicable; i = each of the pollution parameters established by Article 72a 4 (Government of Catalonia, 2003a); n = each type of discharge from the establishment; Rpn = the weighting ratio, based on the flow rate, of each type of discharge in relation to the total types of discharge from the establishment; Ci = the concentration of each pollution parameter for discharged water; Ei = the concentration of each pollution parameter established for incoming water. If Ei is greater than Ci, the value of the difference ($Ci - Ei$) is 0; Pui = the unit price for each pollution parameter established; Cpi = the peak coefficient for each parameter expressing the ratio between the maximum and mean pollution concentration values; Ks = the salinity coefficient; Kdi = the

Table 5.4 Pollution parameters and pricing tax.

Pollution parameters	**Parameter Prices**
Suspended solids	€0.5011/kg
Oxidizable matter	€1.0023/kg
Soluble salts	€8.0198 Sm^3/cm
Inhibitory substances	€11.8889 Kequitox
Nitrogen (N)	€0.7610/kg
Phosphorus (P)	€1.5222/kg

dilution coefficient; *Kan* = the coefficient of discharge to the system for each type of discharge; *Fn* = the fertirrigation coefficient for each type of discharge; *Kr* = the volume correction coefficient that expresses the ratio between the volume of water discharged from and supplied to the entire establishment; and *Kn* = the nutrient coefficient.

5.3.1.10 Tax rate for water supply companies

An exception to industrial and similar uses are water uses in activities related to the collection, treatment and distribution of water IDESCAT (2009), which, as Act 5/2012 of March 20 (Government of Catalonia, 2012), became contributors to the water tax when the exemption of urban water supply activities was lifted. The water tax structure for urban water supply activities is similar to other economic activities; it is composed of a general tax rate according to use and a specific tax rate according to pollution. The specific tax rate is subject to a coefficient of 0, and the general tax rate is affected by the following coefficients: a coefficient of 0.07 if the volume of water used and supplied comes directly from the environment or from infrastructure belonging to the Catalan Water Agency or a public corporation belonging to the water authority, or is produced by a sea water treatment facility; a coefficient of 0.2 for the volume of water used and not billed ("lost water").

On the one hand, the supply-based water tax implies an extra cost for water companies, who have to recover it through the price; however, it has served as an incentive to boost policies to improve and upgrade water supply networks. It has encouraged some companies and municipalities to upgrade some of their facilities. The following graph (Figure 5.6) shows how the cost of the water tax increases according to the performance of the supply system.

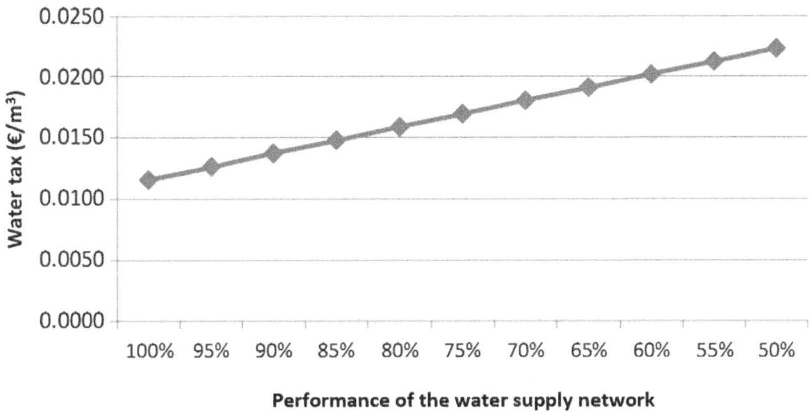

Figure 5.6 Tax rate for water supply (€/m^3) for 2017.

5.3.1.11 Sewage tax

Municipal sewage networks require maintenance, cleaning and conservation to respond to the requirements of citizens, who demand good public services and a high-quality urban environment, and to the growing environmental responsibility and the resulting conformity with legislation. The actions carried out are usually financed through the sewage tax, although not all municipalities apply it.

Funds raised from the sewage tax are used for the comprehensive, orderly and responsible management of the sewage network and urban drainage, with a special focus on preventive maintenance. The inspection and preventive maintenance plan includes everything from cleaning of the sewage network, inlets and grates, to removal of obstructions and urgent cleaning tasks. It also includes waste management and control of discharges. The predominant structures are characterized by a fixed charge with no variable items (26% of municipalities), or with just one consumption block (49%) (Table 5.5). It is worth noting that in some municipalities, the basis for calculating the sewage tax is the cadastral value of the home, regardless of the amount of water consumed.

5.3.1.12 Distribution and dispersion of the domestic tariff

The cost of the drinking water supply differs greatly between municipalities due to determining technical variables, such as the source of the water supply, the type of

Table 5.5 Variable component of sewage network.

Structure of the Variable Component of the Sewage Network	Municipalities with > 5,000 Inhabitants (%)	Inhabitants of mun. > 5,000 Inhabitants (%)	Municipalities Catalonia (%)	Inhabitants Catalonia (%)
Zero consumption blocks	27.46	12.51	61.49	17.82
One consumption block	48.59	35.28	28.15	34.70
Two consumption blocks	7.04	40.52	4.84	36.70
Three consumption blocks	10.56	7.84	3.28	7.18
Four consumption blocks	4.23	2.85	1.55	2.63
More than four consumption blocks	2.11	1.00	0.69	0.97

Source: ACA (2017).

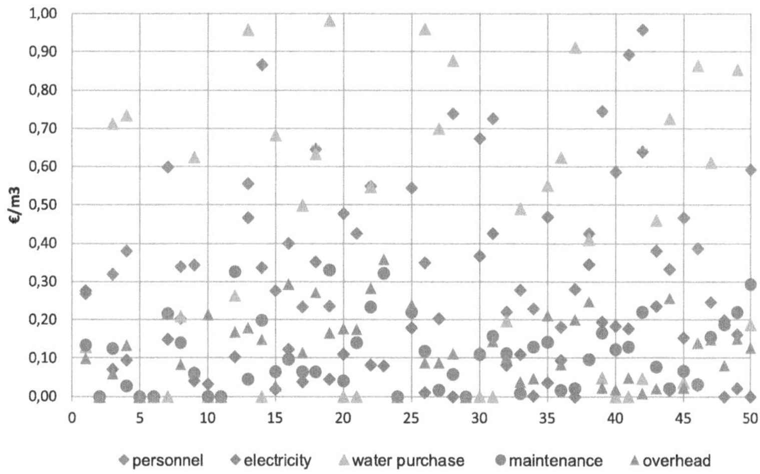

Figure 5.7 Operation costs of different water supply companies (€/m^3).

treatment used, seasonal variations in the municipality, the length of the supply network, the altitude of the municipality and the quality and reliability of the service. Economic and financial factors are also influenced by debt levels, the impact of amortization on the tariff, investments made and subsidies received. Figure 5.7 illustrates the heterogeneity of supply company costs, based on a sample of 50 municipalities. It shows, in particular, costs associated with personnel, maintenance, electricity, purchase of water and overheads.

All this variation in costs between water companies and municipalities means that water tariffs in Catalonia vary widely. The differences in costs between supply companies and local competition concerning tariffs are the principal reasons behind the price dispersion in Catalonia (in municipalities and water companies). There may even be different prices within the same municipality if there are several water companies responsible for the domestic water supply. The price differential is as high as 350% in some cases, i.e., more than €400 in the annual water bill. Figure 5.8 shows the distribution of the total water price, including all items, among municipalities with more than 5,000 inhabitants. There are municipalities where the final price per cubic meter is under €1.20, and others where the price exceeds €3.

Figure 5.9 also shows how the disparity increases with lower consumption levels, given that the fixed portion of the price is the one that differs most between municipalities.

5.3.1.13 Average water price

The average price is nothing more than a simplification of a much more complex reality. The different studies conducted use this value to analyze other items, such

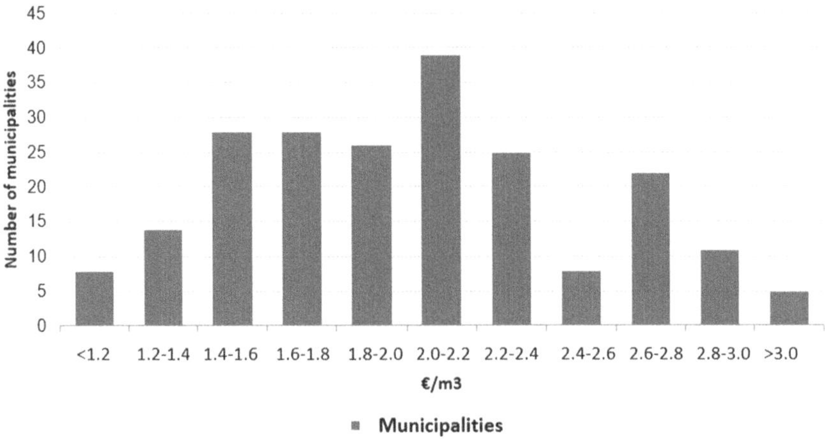

Figure 5.8 Distribution of total water price (VAT included) of the municipalities with more than 5,000 inhabitants (based on consumption of 12 m^3/month, which includes all parts of the cycle: supply, water tax, sewage and VAT).

as tariff progressivity, the weight of each component, and price variability. Thus, it was considered necessary to include this section, since it provides an insight into the behavior of water prices for different household consumption levels.

The provincial prices included here (Figure 5.10) are averages weighted by the number of inhabitants in the municipality; therefore, municipalities with a higher number of inhabitants have greater weight. The results of the study for different monthly consumption levels per household are also shown for 6, 9, 10, 12, 15 and 20 m^3 (Table 5.6).

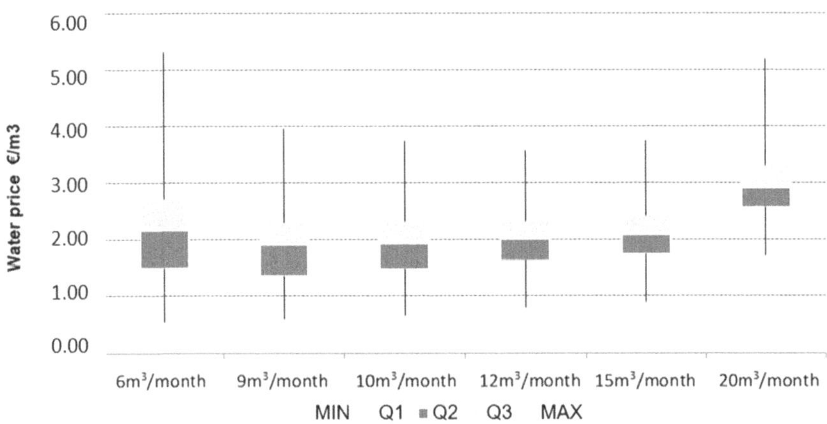

Figure 5.9 Box plot of the water price (VAT included) according to water consumption per month.

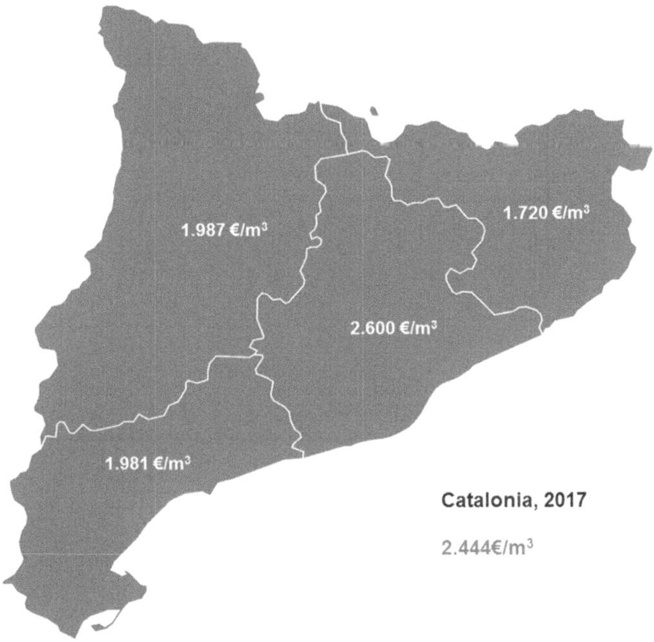

Figure 5.10 Water price average for domestic use for water consumption of 12 m^3/month (VAT included).

The progression of the end price is profoundly shaped by that of the water tax, since the elements that influence it do not have such a strong effect, as illustrated in Figure 5.11, which clearly show the impact that the different components have on the progression of the total water price.

The proportion of the variable component in the water price rises as consumption increases, as reflected in Figure 5.12. In consumption levels of 6 m^3 per month, the variability is just 25%, compared to the variability of 88% for consumption levels of more than 20 m^3 per month.

5.3.1.14 The water bill as a percentage of the household budget

The water bill accounts for 1% of the average household income in Catalonia. This percentage was calculated based on the water bill for a typical household of three members, with a monthly consumption level of 12 m^3 and the average household income €2,611 (INE, 2017). However, it is important to note that the percentage spent on water by a family at risk of social exclusion, with an income equivalent to the Catalan Adequate Income, may be as much as 4% (Figure 5.13). To address this situation, social and economic protection measures have been put in place for the needy.

Table 5.6 Monthly consumption levels of provinces.

Province	Indicator	6 m^3 per Month	9 m^3 per Month	10 m^3 per Month	12 m^3 per Month	15 m^3 per Month	20 m^3 per Month
Barcelona	€/m^3	2.808	2.457	2.507	2.600	2.710	3.664
	€/month	16.85	22.11	25.07	31.2	40.65	73.26
	€/year	202.20	265.32	300.84	374.40	487.80	879.12
Girona	€/m^3	1.925	1.615	1.629	1.720	1.854	2.666
	€/month	11.55	14.53	16.29	20.64	27.8	53.3
	€/year	138.60	174.36	195.48	247.68	333.60	639.60
Lleida	€/m^3	2.261	1.935	1.952	1.987	2.029	2.798
	€/month	13.57	17.41	19.52	23.85	30.44	55.95
	€/year	162.84	208.92	234.24	286.20	365.28	671.40
Tarragona	€/m^3	2.363	1.955	1.955	1.981	2.060	2.978
	€/month	14.18	17.59	19.55	23.77	30.9	59.54
	€/year	170.16	211.08	234.60	285.24	370.80	714.48
Catalonia	€/m^3	2.676	2.320	2.360	2.444	2.547	3.479
	€/month	16.05	20.87	23.6	29.31	38.22	69.54
	€/year	192.60	250.44	283.20	351.72	458.64	834.48

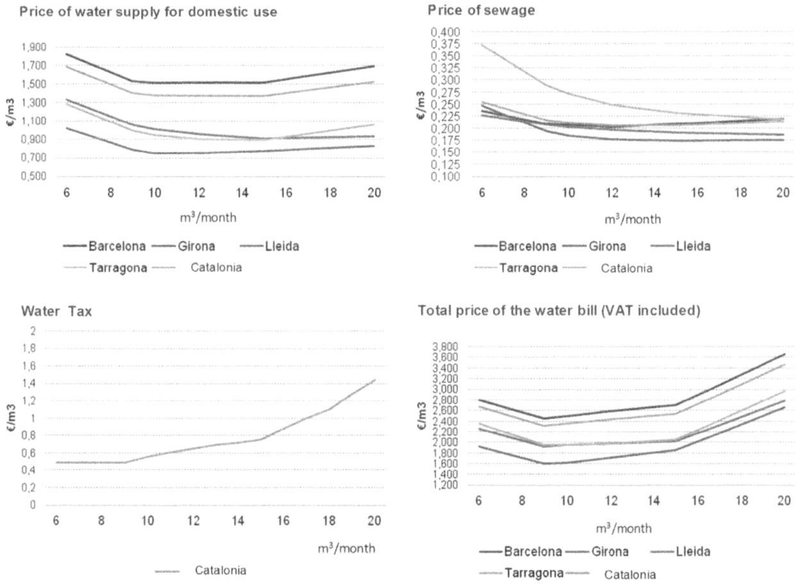

Figure 5.11 Effects on progression of end price of water tax bill.

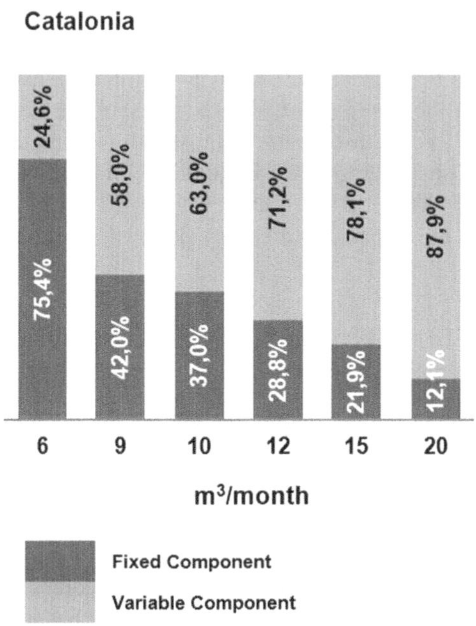

Figure 5.12 Proportion of variable component with consumption levels.

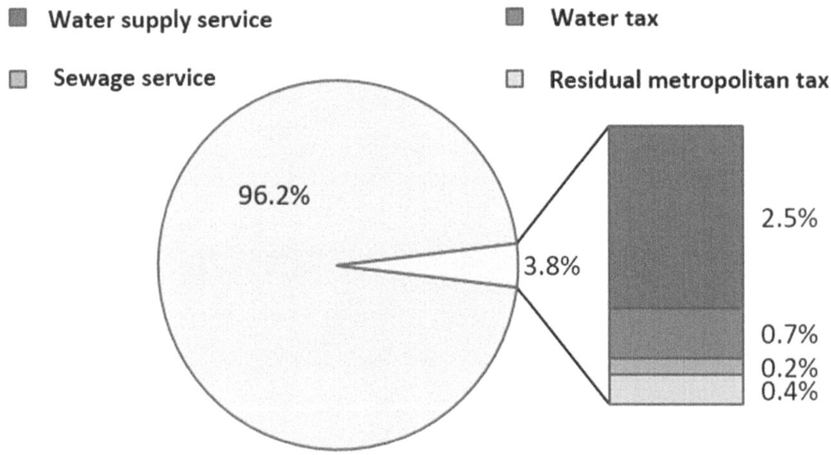

Figure 5.13 Water bill in Barcelona as a percentage in relation to household income.

5.3.1.15 Social tariffs: financial vulnerability

At the European level, Directives 2009/72/EC (European Community, 2009a, b), 2009/73/EC (European Community, 2009a) and 2012/27/EU (European Union, 2012) oblige member states to adopt adequate measures to guarantee water supply to vulnerable households by making it available at affordable prices, especially at critical times of the year, and fighting against energy poverty.

The Catalan parliament urged the Catalan government to take measures to tackle energy poverty with the approval of Motion 56/X of the Catalan parliament of November 7, 2013, which subsequently approved the amendment of Act 22/2010 of July 20 on the Consumer Code of Catalonia (Act 20/2014 of December 29) (Government of Catalonia, 2014b) and Act 24/2015 of July 26 on urgent measures to address the housing and energy poverty emergency (Government of Catalonia, 2015).

Specifically, the government is obliged, in the event of vulnerability, to ensure that spending on the main residence and essential supplies do not exceed the maximum expenditure threshold, equivalent to 30% of the household disposable income (provided that rental and utility expenses are lower than the maximum levels set out in the regulations based on the geographical area of residence of the beneficiary persons or households). In 2015, the Catalan government allocated a social assistance fund for households in situations of economic vulnerability lacking the means to meet payment commitments for water, gas and electricity.

Municipal councils, as the owners of the domestic water supply service, can assist these households by applying social tariffs, discounts and exemptions on the payment of municipal prices and fees, provided that certain conditions or requirements are met. Some municipalities set social tariffs to reduce the economic burden of paying for domestic water services faced by specific groups such as people with low incomes (due to retirement, unemployment, etc.) and those with disabilities.

Social tariffs involve reducing the price to be paid, in this case on the water bill, with the aim of protecting economically and socially vulnerable groups and guaranteeing them access to a certain basic consumption level. Another social measure already being applied by a number of supply companies and authorities is the solidarity fund. Such funds differ from social tariffs in that they are used to pay for all or part of the water bill of vulnerable households without the means to pay the bill, even after reductions have been applied.

In October 2011, the Catalan Water Agency (ACA) approved a social tax to guarantee access to a basic water consumption level and help meet the basic needs of vulnerable households. The social water tax is designed to guarantee that basic water needs are met. Its value is tailored to the situations of vulnerability faced by the neediest members of society. The Social Water Tax consists of applying a special rate for the first water tax block (50% of the usual value), provided that the consumption level of the bill does not exceed this block. The tariff for the first social water tax block is €0.24/m^3 (effective in 2017).

At present, individuals named on the domestic water supply contract (for the residence at which they are registered) and owners of water sources are eligible for the social tax, under the following circumstances: Applicants who are part of a household whose members are all unemployed; Persons in receipt of a minimum contributory pension or compulsory old age and disability insurance (SOVI) due to retirement or widowhood; Persons in receipt of a minimum contributory pension due to a permanent disability; Persons in receipt of a non-contributory pension due to retirement, disability retirement and disability; Persons in receipt of the minimum income allowance (RMI); Persons in receipt of social welfare benefits to cover certain basic needs (Government of Catalonia, 2006); Persons in receipt of benefits from the Social Welfare Fund (FAS); Persons in receipt of benefits from the General Disabilities Act (LGD); Individuals and households that can prove to the supply company that they are financially vulnerable or at risk of residential exclusion by means of a report from the social services department of the appropriate local authority (this is processed directly with the water supply company).

Since the discount is designed to guarantee access to basic consumption, the consumption level may not exceed the limit of the first water tax block, i.e., more than 9 m^3 per month (blocks are extended if more than three people live in the household). In the event that the consumption level is higher, the social tax will not be applied to the corresponding bill; however, this does not make the user ineligible for the tax in subsequent bills. Since the vulnerability of households is not determined by their consumption, but rather by their circumstances and resources, the ACA is drafting an amendment to the social tax so that, without losing the savings incentives of the tax, consumers who are in vulnerable situations but have higher consumption levels can also benefit from the tax.

Municipal councils, as the owners of the domestic water supply service, and water companies, in cases of indirect supply, can reduce the economic burden of paying for domestic water services faced by specific groups by providing non-repayable grants and through social tariffs, discounts and exemptions on the payment of prices and fees, provided that certain conditions or requirements are met.

The social tariff types applied by water companies or councils vary; some apply a discount percentage to all price components (fixed and variable), while others have a specific rate for different groups or a subsidized price for low consumption levels. Non-repayable grants come from the budgets of the council or local authority, or from a solidarity fund created by the company in conjunction with the council or the Catalan government, etc.

Most of the municipalities in Catalonia have social tariffs or funds to cover the water bills of groups in situations of vulnerability. If we limit our analysis to municipalities with more than 5,000 inhabitants, the percentage rises to 57%. Consequently, in 123 of the 214 municipalities with more than 5,000 inhabitants, social discounts are available to 81% of the population.

5.3.1.16 Regulation of water prices

As mentioned above, once the tariffs to be applied to end users have been approved by the relevant local authority (tariff and pricing structures), the proposal is submitted to the Catalan Pricing Commission together with an economic analysis and any supporting documentation. The Pricing Commission is the body that authorizes maximum prices in the field of water supply services for populations, as part of the powers conferred on it by Decree 121/2014 of August 26 (Government of Catalonia, 2014c).

Only 31% of water companies (including councils that are directly responsible for management) request authorization from the Pricing Commission for the tariffs and fees to be applied in municipalities. However, the water supply tariffs authorized by this body affect 88% of the population. (Table 5.7).

Table 5.7 Tariffs authorized by Pricing Commission.

Year	Total Company-Municipality*	Volume Recorded	Subscribers	Permanent Population
Data to December 2016	484	413,001,657	3,389,300	6,600,877
Relative amount for the total for Catalonia (%)	31	86	84	88

*Company-municipality: the unit made up of the supply company and the municipality in receipt of the supply. *Source*: ACA (2016).

Act 2/2014 (Government of Catalonia, 2014a) assigned the function of approving wholesale tariffs to the ACA. This refers to the water supply tariffs of major water companies (wholesale operators) that sell water to other water companies. It is therefore the responsibility of the Catalan Water Agency's board of directors to approve the tariffs for wholesale water supply services.

The role of the regulator in the Spanish water sector is a subject of much discussion, although its functions are clearly to protect user interests, ensure continuity of service, encourage efficiency and innovation, and promote effective competition. It is therefore important to promote and guarantee fulfillment of the following criteria when calculating water supply service tariffs: Transparency: the basic components of the tariff proposals must be explicit and publicly accessible; Financial and economic balance of the service: the tariffs must make it possible to generate sufficient revenue to cover operating and maintenance expenses and to carry out an efficient activity that is sustainable in the long term; Economic efficiency: tariffs cannot be used to pass on cost overruns associated with inefficient management; Equity: fair and impartial treatment; and Simplicity: tariff formulas must be developed in such a way that they are easy to understand, apply and control. There are many price regulation models, from the most interventionist to the least interventionist. This is currently a subject of debate, and at some point, the matter must be addressed.

5.4 COMPARISON OF WATER TARIFFS IN CATALONIA WITH SOME EUROPEAN CITIES

As expected, the key factors that determine the disparity in water prices between European cities are the different costs related to the availability and quality of the resource; regulatory differences (in terms of quality and environmental protection); and the diverse service management models.

In general, the Mediterranean countries of southern Europe attach greater weight to the fixed portion of the water price and create progressive tariff structures with different consumption blocks, i.e., the price increases with consumption. Northern European countries and regions, on the other hand, attach less weight to the fixed portion and have fewer consumption blocks (or none at all) (Figure 5.14).

In France, the fixed portion of water supply tariffs is affected by an idiosyncrasy arising from the reform of the Water Law of December 2006 (Government of France, 2006), which stipulated that water companies are required to limit the proportion of the fixed part with respect to the variable part.

The scarcity of the resource caused by the rainfall pattern in the Mediterranean region forces cities such as Barcelona and Athens to apply four to five progressive blocks in the tariff structure in order to encourage water efficiency. Another example of water scarcity management can be seen in Madrid (Canal de Isabel II), where a seasonal tariff is applied from June 1 to September 30 in order to encourage the rational use of water during the summer months (Figure 5.15).

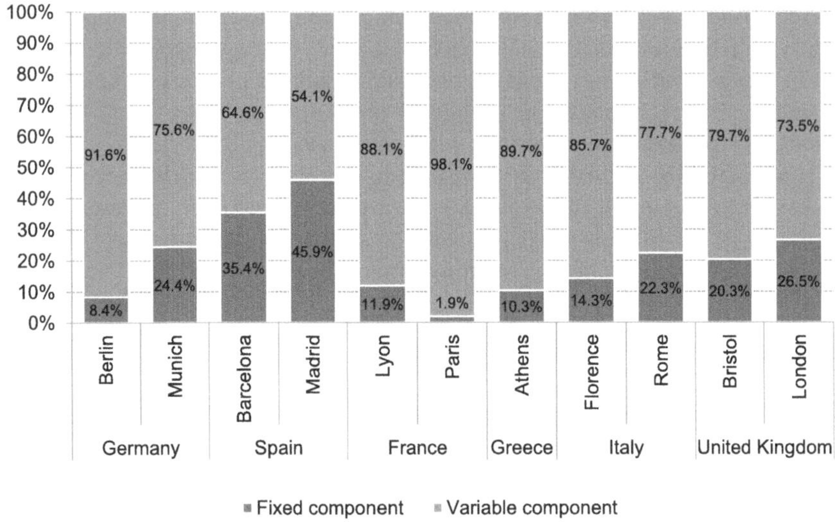

Figure 5.14 Structure of the final water price by country, including fixed and variable components. Water supply and sewage services and water taxes for a domestic use of 10 m^3/month (13 mm meter).

Ultimately, the disparity between the prices in different European cities of services with similar standards requires a case-by-case analysis, and particular attention should be paid to the form and intensity of the impact on environmental and resource costs (Figure 5.16).

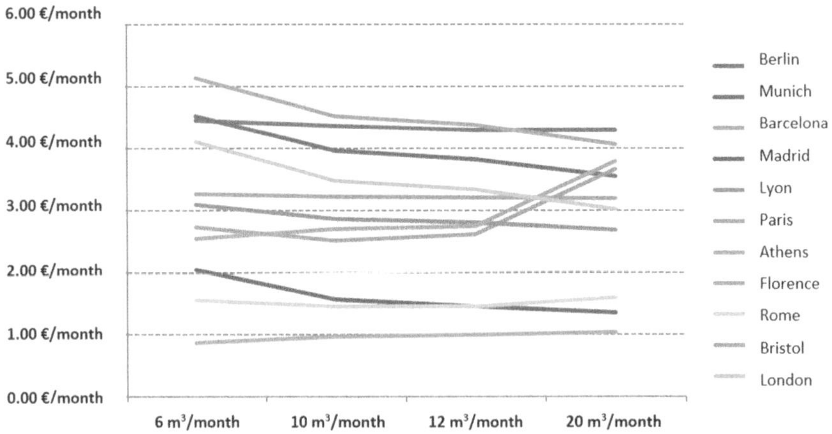

Figure 5.15 Progressive blocks in the tariff structure, and price rate of water supply, sewage and taxes (13 mm meter).

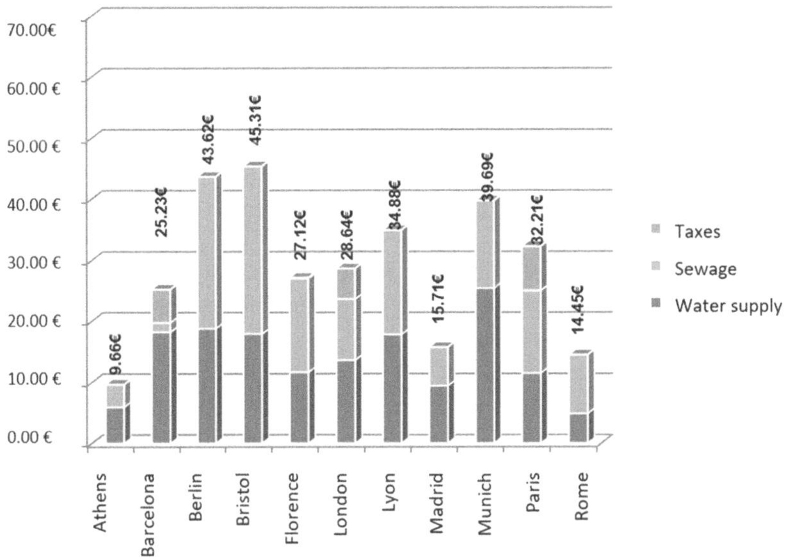

Figure 5.16 Total cost (VAT not included) for water consumption of 10 m^3/month (13 mm meter).

5.5 CONCLUSIONS

The tariff system of the urban integral water cycle, must assume the objective of global self-financing. Water cycle revenues should not be used to cover expenses outside of this cycle and the price must cover all costs incurred. Financial costs to be considered in the fixation of water prices are both OPEX and CAPEX related to water supply and sanitation, in particular: personnel, energy, water purchase, maintenance, remuneration of operators (in the indirect management) and capital costs as technical repayment and others. To sum up, there is a need to internalize at least these costs in order to guarantee the maintenance and reasonable development of the service.

The water canon was established to recover water treatment financial costs but also environmental and resource costs too, homogenizing its impact and structure for the entire Catalan territory, including social measures or guarantee of access to the resource. The lack of investment in municipal hydraulic infrastructures in Catalonia, with a renewal rate of less than 1%, implies that the current infrastructures will have to last 100 years when in practice they are sized for lower periods. This is a risk that should be avoided.

There is a high disparity in the price for the water supply and sewer service. Some municipalities recover full costs on the water bill and others do not, because they do not consider infrastructure and renovation costs. There are different prices due to the technical variables that condition the costs, such as origin of the water, type of treatment, seasonality of the municipality, length of the supply network, quality and guarantee of the service, etc. But these variables do not completely justify such a great disparity of prices. There are cross subsidies and direct and indirect subsidies that help this heterogeneity. Most of the 947 municipalities of Catalonia have a different water price and a great number of them have more than a price in the same municipality, because different nuclei are supplied by different companies.

In summary, the recovery of costs in urban water services is certainly crucial to guarantee the quality of services under the principles of transparency, public information and citizen participation.

REFERENCES

ACA. (2016). Catalan Water Agency. Study on Retail Supply Services in Catalonia.

ACA. (2017). Catalan Water Agency. Water Price in Catalonia at 2017. (In Catalan).

French Constitutional Court. (2016). Sapin II Law: on transparency, anti-corruption and modernization of the economy. *JORF*, **287**, 10.12.2016. (In French).

Government of Catalonia. (2003a). Legislative Decree approving the consolidated text of the municipal and local regimen of Catalonia of 28 April. *DOGC*, **3887**, 20.05.2003. (In Spanish).

Government of Catalonia. (2003b). Legislative Decree approving the consolidated text of the legislation relating to water in Catalonia (TRLAC) of 4 November. *DOGC*, **4015**, 21.11.2003. (In Spanish).

Government of Catalonia. (2006). Act 13/2006. Social welfare benefits to cover certain basic needs. *DOGC*, **4691**, 04.08.2006. (In Catalan).

Government of Catalonia. (2012). Act 5/2012 of March 20 on fiscal, financial and administrative measures. *DOGC*, **6094**, 23.03.2012. (In Spanish).

Government of Catalonia. (2014a). Act 2/2014 on Fiscal, Administrative, Financial and Public Sector Measures of 27 January. *BOE*, **69**, 21.03.2014. (In Spanish).

Government of Catalonia. (2014b). Law 20/2014, of December 29, amending the Law 22/2010, of July 20, of the consumer code of Catalonia, improving the protection of people in the field of consumer credit and mortgage loans, economic vulnerability and consumer relations. *DOGC*, **6780**, 31.12.2014. (In Spanish).

Government of Catalonia. (2014c). Decree 121/2014 on Catalan Pricing Commission of 26 of August. *DOGC*, **6696**, 29.08.2014. (In Catalan).

Government of Catalonia. (2015). Act 24/2015 of July 26 on urgent measures to address the housing and energy poverty emergency. *BOE*, 216 09.11.2015. (In Spanish).

Government of France. (2006). Law no 2006-1772 of December 30th, 2006 on the water and the aquatic environments. *JORF*, **303**, 31.12.2006. (In French).

IDESCAT (2009). Statistical Institute of Catalonia. Catalan Classification of Economic Activities (CCAE). https://www.idescat.cat/Classif/?TC=4&V0=1&V1=87

INE. (2017). Spanish Statistical Office. Life conditions. http://www.ine.es/dyngs/INEbase/categoria.htm?c=Estadistica_P&cid=1254735976608

EUROPEAN DIRECTIVES

European Community. (2000). Water Framework Directive 2000/60/EC of the European Parliament and of the Council of 23 October 2000, establishing a framework for Community action in the field of water policy. *Official Journal*, **L327**, 22.12.2000.

European Community. (2009a). Directive 2009/72/EC of the European Parliament and of the Council of 13 July 2009 concerning common rules for the internal market in electricity and repealing Directive 2003/54/EC. *Official Journal*, **L211**, 55.

European Community. (2009b). Directive 2009/73/EC of the European Parliament and of the Council of 13 July 2009 concerning common rules for the internal market in natural gas and repealing Directive 2003/55/EC. *Official Journal*, **L211**, 14.8.2009.

European Union. (2012). Directive 2012/27/EU of the European Parliament and of the Council of 25 October 2012 on energy efficiency, amending Directives 2009/125/EC and 2010/30/EU and repealing Directives 2004/8/EC and 2006/32/EC. *Official Journal*, **L315**, 14.11.2012.

Chapter 6

Urban water regulation in Europe

Mónica Maldonado-Devis and*
Vicent Hernández-Chover

Water Economics Group, Department of Applied Economics II, University of Valencia, Avda. Tarongers s/n, 46022 Valencia, Spain
**Corresponding author Email: Monica.Maldonado@uv.es*

ABSTRACT

The methods of managing water resources have modified over time, as has the legal framework that sustains it. Traditionally management was based on modifying the supply to satisfy all demands, but has now become more focused on the management of the demand. New holistic approaches to the management of water resources have led to the predominance of certain economic instruments to address water problems. It is not possible to talk about the legal regime of the urban water cycle in isolation. The water resources that are used to supply cities are a percentage of the total available in a certain area and, therefore, are included in a delimitation and planning process that exceeds the urban scope. Water management in general is subject to a legal and administrative framework, which in many cases is exceedingly complex, setting out who intervenes in the organisation of resources, how they are managed and the competences assigned to each administration.

6.1 INTERNATIONAL REGULATION FRAMEWORK

Concern for the environment at the international level first acquired a more organised character after the Second World War, although it was not until the

Editor(s): Francesc Hernández-Sancho
doi: 10.2166/9781780404684_0103

publication of the first Meadows Report in 1972, "*The Limits of Growth*", when the international community began to question the current model. Although it was held at the same time of the publication of the Meadows Report, the United Nations Conference for the Human Environment, also known as the Stockholm Conference, constituted the first international forum where the development model of Western society was questioned. The environment thus became for the first time an issue of international importance in which the Stockholm Declaration was established; this maxim governs the protection and conservation of the environment as one of its basic principles: "the polluter pays".

During this first stage, the issue of water resources was not considered as a main issue, although it occupied a growing space in the field of environmental policy at the international level. An example of this is the Action Plan approved at the first United Nations Conference on Water held in 1977 in Mar de Plata, which focused on the management of water resources as a determining factor in the improvement of economic and social conditions. As a consequence of this approach, on November 10, 1980, the General Assembly of the United Nations decided, through Resolution 35/18, to proclaim 1980–1990 to be the International Decade for the Supply of Drinking Water and Environmental Sanitation, during which Member States should commit to achieving universal access to both services.

In 1987, the Brundtland Report was presented to the United Nations Assembly proposing environmental strategies coherent with sustainable development defined for the first time as "one that meets the needs of present generations without compromising the ability of future generations to satisfy your own needs". As a result of the Brundtland Report, the General Assembly of the United Nations convened the second United Nations Conference on Environment and Development (UNCED), better known as the Earth Summit, held in Rio de Janeiro in June 1992. It was a decisive moment in the international negotiations on issues of the environment, general development and, particularly, on water resources. While in the 1980s the debate on water resources focused mainly on the consideration of water as a complement to public health, from this moment the debate was broadened and considered the management and use of water as part of environmental protection and sustainable development.

Among the non-binding legal instruments resulting from the Rio Conference is the so-called Agenda 21, an Action Plan for the achievement of sustainable development that includes a chapter on water resources. The inclusion of water in the documents resulting from the conference reflects the changes in the focus on the use and management of water that began to prevail from this moment. This change in focus in the management of water resources was conceived in the period prior to the Rio Summit with the holding of two international meetings: The Informal Consultation on the Management and Development of Integrated Water Resources held in Copenhagen in November 1991 and the Conference on Water and the Environment held in Dublin in January 1992. In the latter, the so-called "Dublin Principles" were approved in which, for the first time, water was recognised as an essential finite resource for sustainable development and as an essentially economic good. The Dublin principles were included in Chapter 18 of the Agenda 21 action programme and became a catalyst for further action.

These principles were subsequently approved when, in 1994, the Dutch government organised the Ministerial Conference on the provision of drinking water and sanitation with the aim of promoting the implementation of the water-related approaches embodied in Chapter 18 of Agenda 21. In the Ministerial declaration that was finalised at the end of the Conference, the text reaffirms the objective of advancing the universal coverage of supply and sanitation of water, but also includes support for the integral management of water resources (Biswas, 2004).

The World Water Partnership (WWP), initially sponsored by the World Bank, the Swedish International Cooperation Agency and the United Nations Development Programme (UNDP), has as its main objective the promotion of the Integrated Management of Water Resources (IWRM), understood as the coordinated management of water, land and resources with the objective of maximising economic and social well-being without compromising the sustainability of ecosystems Solanes and González-Villareal (2001).

For its part, on the European level on June 16, 1968, the European Water Charter was adopted, which sets out the basic principles for the protection of water resources. From 1973 onwards, the Action Programmes for the protection of the environment of the European communities took place, with the aim of establishing a joint environmental policy (De Gatta Sánchez 2013a,b; Piñeiro, 1989). Through the different programmes of action, an evolution is observed in the treatment of general environmental problems and of water resources in particular, which goes from a problem-correction approach in the first two programmes prior to 1981 to the prevention of them (Álvarez Fernández, 2004).

In implementing the Fifth Action Programme (1993–2000), different directives are adopted that focus on the control of pollution and move away from the quality and management of water resources. However, in 1995 the European Commission asked the EEA to carry out a report evaluating the state of the EU environment which concluded the need to intensify environmental policies within the European framework. Subsequently, in 1996 the European Commission, in "Water policy in the European Community", specifies some of the principles that should inspire future European water policy; this will materialise in the approval of the well known Water Framework Directive (WFD).

6.2 THE WATER FRAMEWORK DIRECTIVE

The entry into force of Directive. 2000/60/EC of the European Parliament and of the Council, 23 October 2000, establishing a community framework for action in the field of water policy, promotes a definitive change in the guidance on how to manage water resources in Europe. The WFD establishes, in contrast to the traditional models of supply development, a new model of water use consistent with the concept of sustainable development in force at the international level (AEAS, 2011).

The WFD itself acknowledges in its recitals that community waters are subject to increasing pressures caused by the continuous growth of demand for good quality water for all uses. The Directive confirms the need to take qualitative and quantitative measures to protect community waters by developing an integrated community water policy. The objective, set out in Article 1 of the WFD, is the establishment of an action framework for the protection of continental surface waters, transitional waters, coastal waters and groundwater. This general objective is specified in Article 4 in a series of environmental objectives for different bodies of water. In order to achieve these objectives, the WFD will establish the administrative division of water resources in hydrographic demarcations and the need to carry out an economic analysis for each demarcation that must be considered when preparing the programme of measures necessary to maintain the good ecological status of all water bodies.

Related to the economic analysis and the 'polluter pays' principle is Article 9 of the WFD, which incorporates the principle of the cost recovery of all water-related services, acquires special importance for the management of water resources, including environmental costs. To meet this objective, Member States must ensure that the water pricing policy encourages the efficient use of resources and that all water uses contribute to the recovery of these costs. The implementation of the WFD poses major challenges for the EU and for each of the member states that must transpose it into their national legislation.

The implementation of the directive arises in several phases. The first of these includes its transposition to the Member States and the identification of the different demarcations and their competent authorities, as well as the environmental and economic analysis of the demarcations. The rest of the implementation phases are divided into six-year cycles, the first cycle being from 2009 to 2015, and at the end of which the compliance with the environmental objectives established in the WFD must be evaluated.

The implementation schedule of the WFD, together with the articles of the text to which it refers, are detailed in Table 6.1. This whole process involves important challenges both for the EU and for each of its Member States, who must transpose it to their national legislations. With the aim of directing and coordinating the actions of the Member States and of the international demarcations, the EU establishes a Common Implementation Strategy (CIS) that materialises in the creation of different working groups that monitor the process.

The legal and administrative regime of water in Spain has been influenced by the evolution in the treatment of water resources at the international level and was clearly modified by the approval, subsequent transposition and implementation of the WFD. With the approval of the Water Law of 1985 (LAW. 29/1985), the

Table 6.1 Calendar and implementation deadlines of the WFD.

Year	Act	Reference Article in the Directive
2000	Entry into force of the Directive.	Article 25
2003	Transposition to national legislations Identification of the Hydrographic Demarcations and competent authorities.	Article 23 Article 3
2004	Characterization of the Demarcations: pressures, impacts and economic analysis.	Article 5
2006	Establishment of monitoring networks. Start of public consultation.	Articles 8 & Annexe V Articles 4 & 15
2008	Presentation of the draft Management Plan for the Demarcation (Basin Plan).	Article 13 & 14
2009	Finalization of Basin plan including Measurement Program.	Articles 13 & 11
2010	Introduction pricing policies.	Article 9
2012	Start-up of the measurement program.	Article 11
2015	Compliance with environmental objectives. End of the first planning cycle.	Article 4
2021	End of the second planning cycle.	Articles 4 & 13
2027	End of the third planning cycle.	Articles 4 & 13

traditional management approach that attempted to regulate the use of resources in response to increasing demands was abandoned and the management of water resources was introduced, together with the idea of protecting the natural resource. The WL of 1985 and its subsequent developments (ROYAL DECREE. 1/2001; LAW. 62/2003) initiated a new treatment of the planning and management of water resources in Spain, thus facilitating the subsequent implementation of the principles contained in the WFD.

The WL provides a normative basis on which the principles adopted in the WFD will be established. On the one hand, all water resources are considered as hydraulic public domains and the hydrographic basin structures are established as indivisible management units (Herráez and Vicens, 2010; ROYAL DECREE. 849/1986; ROYAL DECREE. 650/1987; ROYAL DECREE. 927/1988 and ROYAL DECREE. 775/2015). On the other hand, a hydrographic planning process is initiated that combines aspects of the coordination of water resources reserved for the National Hydrological Plan (LAW. 10/2001 and LAW 11/2005), with aspects of decentralisation that are specified in the Basin Hydrological Plans (ROYAL DECREE. 1664/1998). The obligation to prepare the Basin Plans falls on the Hydrographic Confederations in the case of inter-community basins and in the Basin Organisations of the competent Hydraulic Administrations in the case of intra-community basins.

The Basin Plans must take into account the environmental objectives collected by the WFD, thus incorporating into the process the need to carry out a Strategic Environmental Assessment and a Special Drought Plan. The content of the Basin Plans themselves has been broadened as it is necessary to incorporate the assessment of the pressures and impacts on the water resources of the demarcation, the ecological flows, and nature reserves and protected areas, as well as an economic analysis of the water. Another innovation that is incorporated into the planning process is the obligation to develop, after analysis, a programme of measures to achieve environmental objectives.

Basin organisations evaluate the resources available in the demarcation, along with the estimation of the different uses; additionally, it must by concession or administrative authorisation determine the use of the assets belonging to the Hydraulic Public Domain. In this way, the water authorised by the basin authority for the supply of the urban demand will form part of the urban cycle and must then be managed by competent administrations. Who these authorities are and their competencies in the management of urban water demand will be addressed in the next section.

6.3 ADMINISTRATIVE REGULATION OF THE URBAN WATER CYCLE IN SPAIN

Understanding the operation of the urban water cycle itself and knowing the allocation of the competences of the different administrations involved in the process is of great importance when designing and implementing improvements in the management of the resource. The organisation and allocation of resources of all phases of the cycle, their distribution of powers and the choice of management form have implications for the payments that end users make for the water supply.

This collaboration between the different water administrations makes the distribution system of competences very complex, which in Spain is subject to its supply and sanitation. The existence of shared responsibilities between the different scopes of the regional authorities and town halls, depending on the service and municipality, can lead to different regulations in different geographical areas.

The urban water cycle can be defined as the set of processes by which the water available in nature in its raw state reaches its urban use; this is in contrast to the process by which it is reintegrated to nature and other uses. These processes can be divided into three phases: supply, sanitation and regeneration (AEAS, 2011). The water supply phase includes catchment, purification, high distribution and low distribution, while the second phase, that of sanitation, includes the collection and transport of wastewater and rainwater and its subsequent purification; the sanitation phase also includes a regeneration phase in which treated water is subjected to additional processes that allow its recycling or reuse (MMA, 2007).

In the supply phase, the surface or underground water is captured from the environment in the places determined by the basin authority. The catchment includes all the tasks necessary to extract the water from the natural environment until its delivery for further treatment or distribution. Subsequently, this water is subjected to the processes of purification necessary for it to be suitable for consumption. Once treated, the water is made available to the distribution network that transports it to the point of entry of the municipality responsible for its distribution, including the tasks necessary to deliver water at consumers' homes and businesses.

Once the water has been used, the sanitation phase begins. The water not consumed by the user, now converted into wastewater, together with the rainwater must be collected and transported from its point of entry into the sewer to the point of treatment or discharge. The treatment comprises the physical, chemical and biological processes to which urban wastewater is subjected (both rainwater and those generated by households, commerce or industry located in cities) to reduce its contamination, leaving it in adequate conditions for its discharge into the receiving channel or for other uses (reuse or recycling). Law 7/1985, of April 2, Regulating the Bases of Local Regime (LRBLR) (modified by Law 57/2003 and Law 27/2013), determines that municipalities assume the competences over water supply services, sewerage and wastewater treatment in the terms provided by the legislation of the state and the regional authorities. It is, therefore, the local administration in collaboration with the state and the regional authorities that must ensure the quality of the water supplied, the sufficiency of the service and the establishment of the tariffs, although the approval of these is done at a higher level.

The role of the regional authorities or supramunicipal entities that assume greater responsibility and competences is increasing in the catchment, purification and distribution phases. And, in general terms, the municipalities retain the competencies for the organisation of the domiciliary distribution of drinking water, that is, the supply from the municipal deposits to the interior installations or pipes for the canalisation of the buildings or what we have denominated as low supply. In the same way, the sewage service is of strict municipal competence and includes the collection of wastewater and rainwater and its evacuation to the general collectors or to the final discharge points.

Regarding the treatment of wastewater, we have seen how the local administration, according to the LRBLR, is assigned the competence. However, urban wastewater treatment services have an impact on a territory that exceeds the strict municipal terms, thus legitimising the intervention of other administrations, especially the regional government, as the competent acting authority.

On the other hand, the approval of Directive. 91/271/CEE, of the Council, of May 21, on the treatment of urban wastewater and its subsequent transposition through Royal Decree-Law. 11/1995, of December 28, on the treatment of urban waste water (modified by Royal Decree. 509/1996 and Royal Decree. 2116/1998) establishes the general applicable rules. These service standards have high maintenance costs that fall on the competent local entities and that require the cooperation of the various public administrations, city councils, regional authorities and general state administration (Molinos-Senante *et al.*, 2012).

The regeneration of water and its subsequent reuse or recycling have not been considered in many cases as a phase of the urban cycle due to its scarce utilisation and its null incidence on the water prices paid by users of the urban supply. Royal Decree (RD) 1620/2007, of December 7, establishes the legal regime for the reuse of treated water. In this RD, the concepts of reuse, reclaimed water and water regeneration systems are defined, and the basic conditions for the reuse of treated water and the procedures to obtain the administrative concession required by the Water Law are established. The concession procedure requires the petitioner to submit a water reuse project and the Basin Agency is responsible for examining the reuse project and determining the compatibility of the request with the Hydrological Basin Plan, considering, among others, the ecological flows. The competence in terms of the concession for the use of reclaimed water rests with the Basin Organisations, although the management of the treated water regeneration systems will fall on the administration that is competent in water treatment.

The main activities that make up the urban water cycle are of local ownership and it is, therefore, the city council who is responsible for establishing the conditions and requirements of the service provision. Law. 57/2003, of December 16, on measures for the modernisation of local government, indicates that public services of local competence can be managed directly or indirectly by means of one of the forms foreseen for the service management contract in Legislative Royal Decree 3/2011 of 14 November, thus approving the revised text of the law on public sector contracts.

In the case of direct management, there is not only the possibility that the service is provided by the local entity itself, but also the creation of a local commercial society whose social capital belongs exclusively to the local entity or a public entity thereof. The creation of autonomous organisations or municipal companies as a way to decentralise the management of public services is the form used by 43% of Spanish municipalities (AEAS, 2010).

In the same way, the law allows various forms of indirect management, the concession and the creation of mixed companies being the most commonly used. The concession is formalised through a public service management contract in which the administration remains the owner, but entrusts its exploitation to a third party, either a natural or legal person, who will manage the service. The counterpart is obtained by the winning company through the tariffs or prices paid by the final user or by the town councils themselves, as appropriate; these type of concessions cannot be held in perpetuity.

The other indirect management formula typically used is the company or joint venture. In joint ventures, the capitals of local corporations and individuals contribute in common to compete in performing public services of the local administration. The other forms of indirect management are much less used in Spain. Among them are consortia and interested management. A consortium represents an association or union of local corporations with various public entities to manage matters of common interest. This is an initiative that small-scale municipalities usually undertake, although there are exceptions regarding size, as in the case of the Bilbao Water Consortium.

The administrations that have the competencies on any of the services related to the urban water cycle also have the competence in setting the price, so that different institutional structures will be reflected in the payments that users make for the service of urban water supply. The fixing of different prices or tax figures depending on the regulation and organisation of the service carried out by the local entity providing the service, whether they be rates, fees, public prices or royalties, will ultimately determine the structure and total volume of the payments to be made by users.

6.4 CONCLUSIONS

In recent decades, the relevance of environmental policies has increased significantly. In addition, the improvement of water management has acquired an important role since ensuring water sustainability is crucial because of its environmental, social and economic implications. This necessarily implies the prevention and protection of the state of water bodies, surface and underground, progressively reducing pollution.

In order to face this challenge, different policies and strategies have been developed at an international level. The various regulations and/or established legal frameworks have experienced substantial changes over time, highlighting their orientation towards the management of water demand driven by some international organizations and, especially in Europe, after the approval of the Water Framework Directive.

The entry into force of Directive. 2000/60/EC of the European Parliament confirms the need to take both qualitative and quantitative measures to protect water resources through the development of an integrated policy. It establishes as objective the good ecological status of all bodies of water, and incorporates the principle of cost recovery of all services related to the urban water cycle, including environmental costs.

The implementation of the Water Framework Directive poses great challenges for the European Union and for each of the member states, since the different economic and environmental measures required have a high impact on the urban water cycle. In the Spanish case, the transposition of this Directive into the internal legal order consolidates the importance of water management from a holistic perspective, incorporating the recovery of the economic, social and environmental costs of the resource.

The application of these new approaches together with the participation of numerous institutions at different territorial levels implies a high degree of legal complexity and requires a clear allocation of responsibilities in the management of water resources. Adequate regulation for each water use and which are the competent institutions and the regulatory framework applicable in each case is essential to ensure efficient management of the resource.

REFERENCES

AEAS. (2010). Supply of Drinking Water and Sanitation in Spain. XI National Survey of Supply, Sanitation and Purification. AEAS, Madrid, Spain.

AEAS. (2011). Guide to Tariffs for Water Supply and Sanitation Services. AEAS, Madrid, Spain.

Álvarez Fernández M. (2004). The Water Supply in Spain. Civitas, Madrid.

Biswas A. K. (2004). Integrated water resources management: a reassessment: a water forum contribution. *Water International*, **29**, 248–256.

De Gatta Sánchez D. (2013a). New stage in the environmental policy of the European Union: the seventh program (2013–2020). *Administrative Update*, **3**, 4.

De Gatta Sánchez D. (2013b). The seventh environmental program of the European Union (2013–2020). *Revista Aragonesa de Administración Pública*, **41**, 71–121.

Herráez P. and Vicens X. M. (2010). Right of Waters for Local Entities. Bayer Hnos, SA, Barcelona, Spain.

MMA. (2007). Prices and Costs of Water Services in Spain. Integrated Report on Cost Recovery of Water Services in Spain. Article 5 and Annex III of the Water Framework Directive. Publications Center General Technical Secretariat of the Ministry of the Environment, Madrid.

Molinos-Senante M., Hernández-Sancho F. and Sala R. (2012). Current status and evolution of sanitation and wastewater treatment in the national and international context. *Anales de Geografía de la Universidad Complutense*, **32**, 69–89.

Piñeiro F. J. G. (1989). Community actions and programs in the field of the environment. *Lurralde: Research and Space*, **12**, 9–14.

Solanes M. and González-Villareal F. (2001). The Dublin Principles Reflected in a Comparative Evaluation of Institutional and Legal Arrangements for Integrated Water Management. GWP, Stockholm, Sweden.

LEGISLATION

European regulation

DIRECTIVE. 91/271/CEE of the Council of 21 of May of 1991, on the Treatment of the Urban Residual Waters. DOCE no. 135 of May 30, 1991.

DIRECTIVE. 2000/60/EC of the European Parliament and of the Council of 23 October 2000, Establishing a Community Framework for Action in the Field of Water Policy. DOUE no. 327, of December 22, 2000.

Spanish regulation

LAW. 7/1985, of April 2, Regulating the Bases of the local regime. BOE no. 80, of April 3, 1985.

LAW. 29/1985, of August 2, of Water. BOE no. 189, of August 8, 1985.

LAW. 10/2001, of July 5, of the National Hydrological Plan. BOE no. 161, of July 06, 2001.

LAW. 57/2003, of December 16, on Measures for the Modernization of Local Government. BOE no. 301, of December 17, 2003.

LAW. 62/2003, of December 30, of Fiscal, Administrative and Social Order Measures. BOE no. 313 of December 31, 2003.

LAW. 11/2005, of June 22, by which Law 10/2001, of July 5, of the National Hydrological Plan is modified. BOE no. 149, of June 23, 2005.

LAW. 27/2013, of December 27, on Rationalization and Sustainability of the Local Administration. BOE no. 312, of December 30.

ROYAL DECREE. 849/1986, of April 11, which Approves the Regulation of the Hydraulic Public Domain that Develops the Preliminary Titles, I, IV, V, VI, VII and VIII of the Water Act 29/1985. BOE no. 103, of April 30, 1986.

ROYAL DECREE. 650/1987, of May 8, by which the Territorial Areas of the Basin Organizations and the Hydrological Plans are Defined. BOE no.122, of May 22, 1987.

ROYAL DECREE. 927/1988, of July 29, by which the Regulation of the Public Administration of water and Hydrological Planning is Approved, in Development of Titles II and III of the Water Law. BOE no. 209, of August 31, 1988.

ROYAL DECREE. 509/1996, of March 15, of Development of the Royal Decree-Law 11/1995, of December 28, by which the Norms Applicable to the Treatment of the Urban Residual Waters are Established. BOE of March 29, 1996.

ROYAL DECREE. 927/1998, of July 29, by which the Regulation of the Public Administration of Water and Hydrological Planning is Approved, in Development of Titles II and III of the Water Law. BOE no. 209, of August 31, 1988.

ROYAL DECREE. 1664/1998, of July 24, by which the Hydrological Plans of Basin. BOE no. 191, of August 11, 1998.

Royal Decree. 2116/1998, of October 2, by which Royal Decree 509/1996, of March 15, is Modified to Develop Royal Decree-law 11/1995, of December 28, by which the Standards Applicable to the Treatment of Urban Waste Water. BOE no. 251 of 20 October 2010.

ROYAL DECREE. 1620/2007, of December 7, which Establishes the Legal Regime for the Reuse of Treated Water. BOE no. 294, of December 8, 2007.

ROYAL DECREE. 775/2015 of August 28, which modifies Royal Decree 650/1987, of May 8, which Defines the Territorial areas of Basin Organizations and Hydrological Plans, and the Royal Decree 125/2007, of 2 February, by which the Territorial Scope of the Hydrographic Demarcations is Fixed. BOE no. 207, of August 29, 2015.

ROYAL DECREE-LAW. 11/1995, of December 28, which Establishes the Regulations Applicable to the Treatment of Urban Wastewater. BOE no. 312, of December 30, 1995.
ROYAL DECREE-LAW. 1/2001, of July 20, which Approves the Revised Text of the Water Law. BOE no. 176, of July 24, 2001.
ROYAL DECREE-LAW. 3/2011, of November 14, which Approves the Revised Text of the Law on Contracts of the Public Sector. BOE no. 276, of 11/16/2011.

Chapter 7

PPPS: Concepts and contractual arrangements in the five continents

Renato Parena

Consultant

Corresponding author Email: renatoparena@yahoo.it

ABSTRACT

Water provision and sanitation services are essential to the well-being of citizens, to public health, economic activities and the protection of the environment, and the body responsible for the organization of these services may decide between different governance models. Governance models most commonly used are direct management, delegation and concession through public-private partnerships, and there are great success stories in this sector, whether public or private management companies. Public and Private Partnerships (PPPs) are forms of cooperation between public entities and the private sector that specifically focus on delivering in a sustainable manner large infrastructures of public interest and/or strategic public services. From the public perspective, the principal focus of Public and Private Partnerships (PPPs) mostly lies on promoting efficiency in public services at affordable prices, through risk sharing and benefiting private sector expertise, jointly with the relief they can provide to the immediate pressure on public finances by fuelling new capital injection. In turn, from the private point of view public sector participation may offer important safeguards for private investors, like the long-term stability of cash flows, and can incorporate important social or environmental benefits into the project. The PPPs, which basically operate on a contractual arrangement basis that may vary consistently with the scope they are focusing on, have worldwide developed in the transport sector, in the area of public buildings and equipment, and in the environmental field as well. The

Editor(s): Francesc Hernández-Sancho
doi: 10.2166/9781780404684_0115

experience varies greatly between sectors and from one country to another; anyway PPPs generally represent a small part of total public services and investments. This chapter will focus on the various aspects of public-private partnerships and describe the different kind and concepts of contractual arrangements mostly adopted in the water field. Additionally, it will provide a panoramic situation of PPPs as they run in 65 countries, in five continents, aiming at glancing the typologies of adopted contracts, the responsibility context and the regulation and pricing control systems so that a couple of key overarching issues can be delivered.

7.1 PPPs IN THE URBAN WATER SECTOR

Providing water and sanitation services has been troubled by a long history of under-pricing, and opposition to full cost recovery principles for basically social reasons. These factors have contributed to the unwillingness of many governments to acknowledge water as a finite natural resource and an economic good that needs a market price reflecting the cost of provision and its true value to society (EU Commission, 2009). The result is that many water systems are often operated inefficiently and that services are unreliable, lacking coverage, regular maintenance and good design such that urban water and wastewater are in critical need of improvement.

Such a need is particularly evident in many developing countries and economies in transition, but also in developed countries, where water sector investments require greater efficiency through better management and the use of new sources (WHO/UNICEF, 2014). However, this target will be difficult to meet with public funds alone, as government budgets have shown decreasing trends throughout time, enhanced by the recent financial crisis. Thus, national and local authorities are increasingly looking to a range of private sector partners to provide access to improved management systems and technical options, as well as private investment funds.

Public Private Partnerships (PPPs) refer to any form of agreement (partnership) between public and private parties aimed at providing effective and innovative ways to deliver infrastructure projects and to provide public services (OECD, 2003). At the same time, PPPs are interesting vehicles for the long-term structural development of infrastructures and services, bringing together distinct advantages of the private sector and the public sector.

Whilst the principal focus of PPPs should bc on promoting efficiency in public services through risk sharing and harnessing private sector expertise, they can also relieve the immediate pressure on public finances by providing an additional source of capital. In turn, public sector participation in a project may offer important safeguards for private investors, in particular the stability of long-term cash flows from public finances, and can incorporate important social or environmental benefits into a project (Parena, 2005).

Nevertheless, PPPs should not be misunderstood as a way of "privatization", where the management and ownership of the water infrastructure are transferred to the private sector. A wide range of approaches for involving the private sector in improving the performance of water and sanitation systems exists. In some cases, PPPs involve the private sector in financing, designing and constructing a water service asset while in other cases it may refer to the management rather than the maintenance of the infrastructures (OECD, 2011b).

7.2 TYPES OF PPPs AGREEMENTS

Public-private partnerships vary in the degree of involvement of the private entity in public infrastructure. A PPP is generally based on a contract or agreement, which outlines the responsibilities of each party and clearly allocates risk. However, it is important to underline that in every PPP option the public authority maintains responsibility for organizing and overseeing public service provision and for ultimately ensuring that public needs are met: the fact that the water sector is a natural local monopoly means that a strong regulatory role is required to ensure that performance standards are achieved and the interests of consumers protected (Mohajeri et al., 2003).

The most frequent contracts adopted in the water sector are: Management, Operation and Maintenance (O&M); Lease and *Affermage*; Concession, Build-Operate-Transfer (BOT), Design-Build-Operate (DBO); and Joint Venture (OECD, 2011a). However, it's important to underline that specific contracts may present an overlap between the above types, and the name given to a particular agreement may not reflect this classic categorization (Brunet, 2014). Care should also be taken to verify whether a specific type of contract is incorporated in the laws of the host country, as in the case of many civil laws where there are strict definitions of "concessions" and "*Affermage*." The key features of each type of contract will be described in the following paragraphs.

7.2.1 Management and 'Operation and Maintenance' contracts (O&M)

At first glance, management contracts commonly involve the private operator so as to improve a public utility's performance by staffing management positions, transferring expertise, and training personnel. Operation and Maintenance (O&M) contracts, on the other hand, require the private operator to develop technical operations of the system. In practice the term 'management contract' has been applied to cover a range of contracts from technical assistance contracts through to full-scale operation and maintenance agreements. The most common features are that the awarding authority engages the contractor to manage a range of activities for a relatively short time period (2–5 years). However, longer-term operation and maintenance agreements are becoming more common in the water and energy sectors where more extensive participation by the private sector through Lease, *Affermage* or concession arrangements in these "essential services" is deemed to be too politically sensitive or not feasible.

Management contracts tend to be task specific and input rather than output focused, while operation and maintenance agreements may have more outputs or performance requirements. Management contracts have, in general, limited potential for improvements in efficiency and performance, although more sophisticated management contracts (often structured as operation and maintenance contracts) may introduce some incentives for efficiency or improved bill collection, by defining performance targets and basing a portion of the remuneration on their fulfilment (and cover longer time periods).

The simplest management contracts involve payment of a fixed fee to the private operator by the awarding authority for performing specific tasks, the remuneration does not depend on the collection of tariffs and the private operator does not typically take on the risk of asset's condition. When management contracts become more performance-based, they may involve the operator taking on more risk, even the risks of asset condition and replacement of more minor components and equipment. The key features of management/O&M contracts may be summarized in Table 7.1.

Table 7.1 Management and O&M contracts.

Object	To manage a range of activities, generally, in the short term (usually for two to five years).
Favoured for	Transitional arrangements to introduce the private sector into managing infrastructure.
Generally adopted	In the water sector and, to a more limited extent, in energy sector.
Obligations	Operator is usually paid a fixed fee to cover its staff and expenses. There may also be a performance-based fee and liquidated damages for failure to achieve performance parameters. Operator may also be required to collect bills on behalf of the public utility and may accept some collection risk in terms of performance standards. In some cases the operator may be required to operate and maintain the assets, sometimes extended to bearing the cost of routine replacement of small, low value parts of equipment (this will require more monitoring to ensure that the outputs are achieved and usually involve higher establishment costs).
Connections	Operating agreements are also usual in relation to BOT/concession arrangements, with the concessionaire sub-contracting the operations phase of the concession to an operator.
Criticality	Contractor usually merely adds a layer of management over the existing public utility structure that normally doesn't transfer any employees: this often causes problems if the staff of the public utility look to the awarding authority for instructions and guidelines. Enforcement of discipline by the private operator may also be difficult.

7.2.2 Lease and *Affermage* contracts

Generally speaking, Leases and *Affermage* contracts require a private operator who pays a fee to the Responsible Body which is responsible for funding infrastructures expansions and/or upgrading to ensure the overall management of the service against a remuneration paid by the users through water tariffs (Parena, 2010). This means that the private operator is responsible for operating and maintaining the utility, but not for financing the investment. When private equity and commercial debt are not available for water supply and sanitation, Leases and *Affermage* contracts allow the Awarding Authority to combine public financing with attracting private efficiency and, in so doing, transferring commercial risk to the private operator (OECD, 2000), in a larger amount than with a management contract, against incentives to perform

While under management contracts the operator receives a fixed fee for his services from the Awarding Authority, under Leases and *Affermage* contracts he charges the consumers with an 'operator fee'. In the case of a Lease, this operator fee is included among the receipts, a portion of which is turned as a Lease fee to the Awarding Authority, as owner of the assets. In the case of an *Affermage*, the

Table 7.2 Lease and *Affermage* contracts.

Object	The Contractor is required to care for the overall management of the service over the medium term (typically between 8 and 15 years).
Favoured for	Lease and *Affermage* allow for the introduction of the private sector into managing infrastructure in the scope of attracting private efficiency and passing on commercial and operational risks.
Generally adopted	In the water sector, given the high potential for improvements in efficiency and performance
Obligations	Lease operator will bear the collection and operational risk. When cost of maintenance and some replacement are passed to the operator, he also takes some degree of asset risk in terms of the performance of the assets. The operator may be involved overseeing capital investment program and/or specific capital works and given a minimum maintenance or replacement provisions towards the end of the contract, so that facilities are handed back in an operational state
Connections	Usual alternative to management and O&M contracts when a more extensive participation by the private sector is politically endorsed.
Criticality	Public Utility Employees are seconded or transferred to the operator. Lease operator requires assurances as to tariff levels and increases over term of Lease and compensation review mechanism if tariff levels do not meet projections. Every 4 or 5 years the review process analyses performance, costs, tariff levels, etc.

operator fee is excluded from the receipts but additionally surcharged and then paid to the Awarding Authority to go towards investments that the same Awarding Authority makes/ has made in the infrastructure. Moreover, under Leases and *Affermage* contracts, the operator tends to bear greater operating risk than under management contracts and to employ the staff directly.

In the case of a Lease, the rental payment to the Authority usually tends to be fixed irrespective of the level of tariff collection that is achieved and so the operator assumes a risk on bill collection and on receipts covering its operating costs. In the case of *Affermage* the operator is assured of its fee (assuming that the receipts are sufficient to cover it) and it is the Authority that assumes the risk of the receipts collected from customers covering its investment commitments. In each case, the Awarding Authority remains responsible for financing which is supposed to come, at least in part, from the rental payment/ surcharge, and managing investment in the assets. Some *Affermage* arrangements also require the operator to design and manage the investment program. The key features of Lease and *Affermage* contracts are summarized in Table 7.2.

7.2.3 Concession, Build-Operate-Transfer (BOT) and Design-Build-Operate (DBO) contracts

A 'Concession' is a long-term contract whereby the operator is responsible for fully managing the service through the use and management of all of a utility's existing assets as well as funding, building and operating new assets. So the private operator assumes all of the risk for the condition of the assets and for new investment, having the responsibility for operation and maintenance of the existing assets and also for financing and managing all required investment.

Concessions typically span a period of 25–30 years (which is deemed a long enough period to depreciate a greater part of major initial investments) and may be granted in relation to existing assets, an existing utility, or for extensive rehabilitation and extension of an existing asset. Asset ownership remains with the Authority and reverts to public ownership at the end of the concession period, including assets purchased by the operator, against a compensation corresponding to the not yet depreciated book value. Under a concession regime the operator typically has a direct relationship with the consumer from whom he directly obtains its revenues.

The concept of a 'concession' was first developed in France where, as with *Affermage*, the framework for the concession is set out in the law which emphasizes the public nature of the arrangement (because the operator has a direct relationship with the consumer who benefits from general protection and safeguards declared in the law). Under French law the *concessionaire* has the obligation to provide continuity of public services, to treat all consumers equally and to adapt the service according to changing needs. In return, the *concessionaire* is protected against new concessions, which would adversely affect his rights. Similar legal frameworks have been incorporated into civil law systems elsewhere.

The legal structure closest comparable with concession is Build-Operate-Transfer (BOT), which is typically used for the purpose of constructing a specific new asset, facility, or system, rather than a whole network (although refurbishment of an existing asset may be involved), in what would traditionally be a public sector project. The public sector grants a private company the right to develop and operate a facility or system for a certain period (the 'Concession Period') and the operator finances, owns and constructs the facility or system and operates it commercially for the concession period, after which the facility is transferred to the authority. The revenues generated from the operation phase do not involve consumers but directly come from the fee charged to the public sector (the utility or the government), which are intended to cover operating costs, maintenance, repayment of debt principal (which represents a significant portion of development and construction costs), financing costs (including interest and fees), and a return for the operator.

BOT is the classic tool for 'project finance' agreements run by a Project Company (that usually is a special purpose vehicle) according to a complex contractual structure in which the public sector (through concession agreements), the private operator (according the O&M agreement), the construction contractor (fulfilling the construction contract), the 'off-take purchaser' (who will purchase the project output because of the off-take purchase agreement) and the financial sponsor (fulfilling the input-supply agreement) are partners.

With a Design-Build-Operate (DBO) Project the public sector owns and finances the construction of new assets while the private operator designs, builds and operates the assets to meet certain agreed outputs, does not assume any financing risk and will typically be paid a sum for the design and build of the plant and then an operating fee for the operating period. DBO is in general simpler than a BOT or Concession contract as there are no financing operations and will typically consist of a civil works contract plus an operating contract, or a section added to the contract covering operations. Key features of Concession, BOT and DBO contracts are summarized in Table 7.3.

7.2.4 Joint Venture arrangements

A Joint Venture (JV) is a contractual arrangement (i.e. BOT, DBO, etc.) generally created to establish a partnership for a specific purpose basically focusing on sharing profits between partners. The basic key issues that must be considered when designing joint-venture arrangements cover questions regarding funding, equity division, corporate governance, asset ownership and management responsibility in connection with the different forms the Joint Venture may be established in practice. JV may even be intended to act as the first step of a privatisation process involving the public owner of an existing share-capital utility to divest to the private sector a number of shares in the utility.

Table 7.3 Concession, BOT, DBO contracts.

Object	The concession contractor fully manages in the long term the service, operating and maintaining all existing assets, and funding, building and operating new infrastructures. The BOT contractor finances, owns and constructs a facility or a system and operates it commercially for the concession period, after which the facility is transferred to the Authority. The DBO contractor designs, builds and operates the assets to meet certain agreed outputs without bearing any financing risk.
Favoured for	Concession, BOT and DBO allow public sector to benefit from potentially great improvements in efficiency and performance over the long term as well as rehabilitate and extend infrastructures in what would traditionally be a public sector project. Involved private operators take on all operational and financial risks while the asset ownership remains with the Authority and reverts to public ownership at the end of the concession period.
Generally adopted	Concession operates in every public services sector while BOT and DBO are generally adopted to implement or refurbish large public infrastructures, (like toll roads, dams, incinerators, gas pipelines etc.) often as a project financing tools.
Obligations	Concessions are focused on outputs (i.e. the delivery of a public service) in accordance with well devised performance standards and customer's requirements, the Awarding Authority is constantly monitoring and periodically reviewing. Less focus on inputs although there may be some requirements regarding frequency of asset renewal and consultation with the awarding authority or regulator on key features as maintenance, serviceability and renewal of assets.
Connections	BOT and Concession arrangements are often in relation with O&M agreements with the concessionaire sub-contracting the operations or the construction phase of the concession to a third part operator.
Criticality	The operator needs assurances that it will be able to cover the costs, finance its obligations and still maintain a profitable rate of return while the Awarding Authority and Regulators need assurances that customers rights be respected at any time in both terms of quality and affordability of the service. In many developing countries there are sectors where the tariffs structures don't respect full cost recovery principles so, in these cases, a clear basis of alternative cost recovery will need be set out in the concession, whether from general subsidies, from taxation or from loans from government or other sources. Appropriate safeguards and grants for periodical reviews of mutual obligations will thus need to be included in the concession agreement or in legislation.

Table 7.4 Joint venture arrangements.

Object	Partners jointly establish an agreement for a specific purpose and basically focus at sharing profits between them. The agreement generally acts over a finite time period and may be targeted to develop a new entity or a new asset by contributing equity. Partners may even enter into a joint venture in a consortium when seeking technological expertise or technical service arrangements, management contracts, rather than rental agreements, for one-time contracts. In this case the JV is dissolved when that goal is reached.
Favoured for	Unlike a merger or acquisition, there is no transfer of ownership from one owner to the other, so owners remain as separate entities. Alliances can be highly flexible, ranging from one specific purpose or project (in the case of a consortium for tendering purposes) to an open-ended spectrum of results (for example, developing a new technology platform faster than the owners could have done on their own rather than running a public service utility in a more effective and efficient way) or for a combination of both purposes and results. JVs typically have a finite life.
Generally adopted	JV operate in every public services sector, even in combination with O&M, BOT and DBO contracts, and are generally adopted to involve the private sector in building new assets or providing public services, and in this last case may advance a wider privatisation process.
Obligations	Under the joint venture model, the public and private sector partners accept the idea of shared risk and shared reward by contributing equity. Reciprocal goals do not have to be identical for the joint venture to succeed, but they must merely be compatible (e.g. the financial return and the improved quality of service) and lead to a common or shared outcome (e.g. the creation of a viable and sustainable organization of the involved public service). The Public partner (normally the local government) is both the ultimate regulator (having the institutional obligation to safeguard public needs and interests and the responsibility for ensuring that adequate and affordable infrastructure services are provided to all citizens), as well as an active shareholder in the operating company (looking for sharing the expected profit) while the private sector partner often has the primary responsibility for managing the day to day operations (into a business concept driven by profit).
Connections	Every other PPP arrangement may be in relation with JV agreements with the consortium or the operating company sub-contracting the operations or the construction phase to a third part operator.

(*Continued*)

Table 7.4 Joint venture arrangements. (*Continued*)

Criticality	The separation of ownership into distinct entities can increase the complexity of the relationship between partners from both the strategic and risk-sharing perspective and the day-to-day operations due to the fact that managing the resources of each respective organization can often conflict with providing the right level of support that the JV needs. The typical flexibility of JVs makes defining a strategy and short- or long-term goals essential to success and the finite life of JVs (even if it is a long period of time for capital-intensive projects) as well as requiring that capital allocation decisions, resource/operational planning and strategic goals must align with the timeframe of the JV.

The level of shared ownership will differ depending on whether the government wishes to retain (at least in an initial period) management control of the utility and its management framework (from holding the majority of the shares in the entity to just keeping a significant influence on the shareholder's decisions). For strategic reasons, mainly connected with the nature of involved public services, the public sector will often keep control of the entity, particularly if the joint venture company owns the assets, while the private sector will avoid any limitation to the company management power (even requiring, in the constitutional documents of the company and the shareholders' agreement, specific veto or weighted voting rights on certain issues).

Typically, the operation and maintenance functions are then delegated to the private operator through a management contract. In the case of a new investment project, public and private partners may establish a project company with a joint share ownership structure or simply agree to bear the separate full legal responsibility for the project without creating a separate legal entity. In other cases, a contractual consortium arrangement in which the parties contract to work together on a specific project, may be established agreeing that one of the involved parties plays the joint mandate and each party is remunerated for the specific services provided to the consortium.

The consortium agreement will thus cover the key issues that need to be established between parties before they start developing a project, or preparing response to tender documentation for a project, rather than creating any separate legal entity or defining concepts of sharing profits. Key features of joint venture arrangements are summarized in Table 7.4.

7.2.5 Breakdown of responsibilities in PPP arrangements

After briefly describing the PPP agreements mostly adopted in the water sector, Table 7.5 holistically and synthetically presents how public and private

Table 7.5 Breakdown of responsibilities.

	Setting Performance Standards	Asset Ownership	Capital Investment	Design and Build	Operation	Fee Collection	Oversight Performance and Fee
Full public provision							
DBO							
Concession							
Joint venture							
BOT							
Full private provision							

Key: grey = public responsibility; grid = shared public/private responsibility; white = private responsibility.
Source: Yale-UNDP Partnership program 1998 and OECD (2000). Global trends in urban water supply and wastewater financing and management: Changing roles for the public and private sector, Paris.

responsibilities are allocated across the different forms of private involvement in the water sector.

7.2.6 England and Wales: a particular experience of full privatisation

Since 1989, England has developed an institutional and regulatory framework for water services regulation and provision based on a delegated private management system, where public assets of the water industry have been completely divested to private companies (Garcia Quesada, 2011). The Water Industry Act and the Water Resource Act established in 1991 are the basis for the new framework for 10 privatized water and sewerage companies, operating within regional monopolies, and 21 companies that provide only water supply in different areas of the country. These firms are in charge of providing water supply and sewerage services under the direction of the national government and national regulators – OFWAT, the Environment Agency (EA) and the Drinking Water Inspectorate (DWI), which are responsible for the adoption and monitoring of environmental and water quality standards of either EU or national origin.

The regulatory framework gives particular prevalence to water consumer associations such as the Consumers Council for Water, which act as an intermediary organisation mediating between regulators and water consumers at large and play an active role as participant in decision-making processes to set water prices and water service standards. At privatisation all water companies became public limited companies, with the core business of providing water services under a licence agreement that allows them to supply in designated areas for 25 years and with the obligation to comply with any direction (including performance standards a price cap for the services that they provide to consumers) given by OFWAT's Director, the Secretary of the State and, in certain cases, by the Competition Commission.

The national government is in charge of all regulation aspects of water policies in the UK, through the Department for Environment, Food and Rural Affairs (DEFRA), and appoints and supervises the activities of the regulators, thus playing a pivotal role in setting both water prices and water quality standards. The Consumer Council for Water (CCWater) represents water and sewerage consumers in England and Wales, operates, as a non-departmental public body, to gather and disseminate information about the functioning of water services companies and reports to DEFRA.

Whereas not directly involved in the regulation of water services provision, the Environmental Agency (EA) participates in the regulation of the water industry securing the proper use of water resources in England and Wales through developing and enforcing water quality standards for inland, estuarial and coastal waters, under the direction of the national government. The Drinking Water Inspectorate (DWI) is part of DEFRA and, as such, is the body in charge of regulating public water supplies in England and Wales and of inspecting water companies, monitoring qualitative standards and investigating consumers' complaints.

7.3 CONTRACTUAL ARRANGEMENTS IN THE FIVE CONTINENTS

The intent of the authors is to be clearly neutral towards any form of governance (totally public or totally private rather than involving both into a participation project). Nonetheless there is considerable evidence that PPPs can: improve delivery of projects in public infrastructures at cross-border level (examples include in the EU Trans-European Transport (TEN-T) network); provide better value for money from infrastructure by exploiting the efficiency and innovative potential of a competitive private sector to either costs, or achieve a better quality ratio (IBNET, 2010); spread the cost of financing the infrastructure over the lifetime of the asset, thus reducing immediate pressures on public sector budgets and allowing the completion of infrastructure projects to be brought forward by a number of years; improve risk sharing between public and private parties so that a proper risk apportionment may induce a more efficient risk management and thus reduce the overall costs of projects; boost sustainability, innovation and research and development efforts; benefit from the private sector experience and resources in developing and implementing long-term strategies for major industrial, commercial and infrastructure programmes; and improve the access to underperformed public services in developing countries through giving large public works and utilities the needed critical mass to tender public offers for BOT projects and service concessions.

PPP experience varies greatly between sectors and from one country to another. Many European states have limited experience of PPPs or none at all. At a global level, the spread of PPPs, in terms of overall management of public services or the construction and operation of public infrastructure, is still very limited and represents a small part of total public investment. Governments have reacted differently to the negative impact, mainly due to significant increases in the cost of debt for PPP projects (induced by the credit crunch) and the reduced maturities offered by banks on their debt, the recent financial crisis had on PPP projects.

Given that investing in infrastructure projects is a key way to maintain the economic activity during crisis periods and support a relatively rapid return to sustained economic growth, some authorities have launched supporting measures in the scope to stabilize the financial sector and limit the impact of the recession on citizens and the real economy while others have reduced, or temporarily suspended, their PPP programmes.

Several studies review PPP state of the art, each of them from different points of view and in different contexts. The following sections present the results of a wider investigation that, taking as its basis the databases of major European private players in the water sector PPP projects and the "International Statistics for Water Services" the IWA Specialist Group on Statistics and Economics periodically updates, overarches the various aspects of PPPs in the water sector as they result in 2012. Data cover 65 countries in the five continents (even if with a non-equitable distribution) and looks at the different contractual arrangements most generally adopted, the responsibility context and the regulation and pricing control systems (IWA, 2004).

7.3.1 Africa

Water and sanitation coverage as well as the percentage of these that are managed under a PPP agreement in African countries are presented in Table 7.6. In general, the majority of water and sanitation services are publically managed. The exception is Gabon where 92% of water services are under a PPP management framework; however, at the same time, all sanitation services are publically managed. Algeria also presents a significant proportion of water services under a PPP arrangement (40%).

The majority of PPP contractual arrangements are BOT and BBO concessions, followed by management contracts and Leases (Table 7.7).

The provision of water services at the national level is a ministerial responsibility (generally the Ministry of Water Resources or the Department of Water Affairs) or that of the Central Government (Gabon, Ivory Coast, Tunisia). Municipalities have a lower level responsibility in Niger, Morocco, Uganda and Zambia while Water Service Boards or Basin Agencies act at an intermediate level of responsibility in

Kenya, Morocco and South Africa. Regulator supervises and monitor pricing policies only in Egypt, Kenya and Zambia. Finally, only Ivory Coast specifically includes in its national legislation full-cost-recovery issues (EU Commission, 2009).

Table 7.6 African countries water services: access and governance.

Country	Population (M Inhab.)	Urban Popul. (%)	Access to Water (%)	Access to Sanitation (%)	Public Management (%)		PPP Management (%)	
					Water	Sanitation	Water	Sanitation
Algeria	38.7	70	77	n.a	60	86	40	14
Egypt	87.5	42	99	50	100	83	0	17
Gabon	1.7	85	92	n.a.	8	100	92	0
Ivory Coast	23.8	45	76	26	78	86	22	14
Kenya	41.8	25	39	15	100	100	0	0
Morocco	33.4	58	96	88	79	85	21	15
Niger	17.1	18	13	n.a.	87	n.a.	13	n.a.
South Africa	54.0	64	78	53	99	100	1	0
Tunisia	11.0	66	83	n.a.	98	100	2	0
Uganda	36.6	15	74	6	98	100	2	0
Zambia	15.0	40	82	56	100	100	0	0
	360.6	**45**	**65**	**42**	**82**	**94**	**18**	**6**

Note: population data refer to official and World Bank estimations for 2013–2014; other data refer to 2012.
Source: World Bank (2014a, b).

Table 7.7 Contractual arrangements of PPPs in Africa.

Contractual Arrangements	Country	Incidence (%)
Management contracts	Algeria, Uganda	33
Operation & Maintenance	South Africa	17
Lease (*Affermage*)	South Africa, Uganda	33
Concession – BOT/DBO	Gabon, Ivory Coast, Morocco, Niger	83
Joint venture	South Africa	<1

7.3.2 Americas

Table 7.8 presents water and sanitation coverage as well as the percentage of these that are managed under a PPP agreement in the Americas (UN Economic Commission for Latin America and the Caribbean, 2014; IMTA/OECD, 2008).

Table 7.8 American countries water services: access and governance.

Country	Population (M Inhab.)	Urban Popul. (%)	Access to Water (%)	Access to Sanitation (%)	Public Management (%)		PPP Management (%)	
					Water	Sanitation	Water	Sanitation
Argentina	42.7	90	87	65	96	96	4	4
Brazil	203.4	84	85	55	69	77	31	23
Canada	35.5	81	36	31	93	94	7	6
Chile	17.8	89	100	96	23	26	77	74
Colombia	47.9	77	94	90	76	89	24	11
Mexico	119.7	79	90	89	87	75	13	25
Peru	30.8	78	57	n.a.	97	90	3	10
U.S.A.	319.1	81	100	46	85	93	15	7
	816.9	**82**	**81**	**59**	**78**	**80**	**22**	**20**

Note: population data refer to official and World Bank estimations for 2013–2014; other data refer to 2012.
Source: World Bank (2014a, b).

The country that has adopted to a very great extent PPP management agreement for water and sanitation provision is Chile (UN Economic Commission for Latin America and the Caribbean, 2014; IMTA/OECD, 2008). Brazil has also adopted PPP management, but in less than half of the proportion of Chile. As is the case in Africa, the majority of PPP agreements are BOT or DBO concessions (Table 7.9).

Table 7.9 Contractual arrangements of PPPs in the Americas.

Contractual Arrangements	Country	Incidence (%)
Management contracts	Brazil, Canada	25
Operation & Maintenance	Brazil, Canada, USA	38
Lease (*Affermage*)	Brazil, Colombia	25
Concession – BOT/DBO	Argentina, Brazil, Colombia, Mexico, Peru	63
Joint venture	Brazil, Chile, Mexico	38

At the national level the provision of water services is a ministerial responsibility (generally the Ministry of the Environment) or that of National Agencies or Authorities (the National Secretariat for water resources in Argentina, the Regulator in Chile or the national Agencies in Mexico, Brazil and USA). Municipalities have a lower level of responsibility in every country while Provinces

Table 7.10 Asian and Oceanian countries water services: access and governance.

Country	Population (M Inhab.)	Urban Popul. (%)	Access to Water (%)	Access to Sanitation (%)	Public Management (%)		PPP Management (%)	
					Water	Sanitation	Water	Sanitation
Australia	23.7	89	100	94	53	81	47	19
China	1,367.9	53	91	56	85	85	15	15
India	1,090.0	32	n.a.	n.a.	99	100	1	0
Indonesia	252.2	52	22	15	94	100	6	0
Iran	77.9	72	93	28	100	100	0	0
Israel	8.3	92	100	99	83	100	17	0
Japan	127.1	92	98	73	100	99	0	1
Jordan	6.7	83	n.a.	n.a.	100	46	0	54
Lebanon	4.1	88	n.a.	n.a.	99	100	1	0
Malaysia	30.4	73	94	n.a.	47	100	53	0
Palestinian	4.5	72	97	58	100	100	0	0
Saudi Arabia	30.7	83	n.a.	n.a.	54	69	46	31
Singapore	5.5	100	100	100	51	100	49	0
South Korea	50.4	82	95	92	100	97	0	3
Thailand	64.9	32	n.a.	n.a.	97	100	3	0
	3,144.3	**49**	**89**	**68**	**84**	**92**	**16**	**8**

Note: population data refer to official and World Bank estimations for 2013–2014; other data refer to 2012.
Source: World Bank (2014a, b).

or Federal Governments present an intermediate level of responsibility in Argentina, Brazil, Canada, Mexico and USA. Regulator supervises and monitor pricing policies in every country examined (excluding Canada) while only in Brazil, Chile and USA the full-cost-recovery issues are specifically included into the legislation (EU Commission, 2009).

7.3.3 Asia and Oceania

Table 7.10 presents water and sanitation coverage as well as the percentage of these that are managed under a PPP agreement in Asia and Oceania. Malaysia, Australia, Singapore and Saudi Arabia have largely adopted PPP management agreements for water services while Jordan and Saudi Arabia have the highest adoption for sanitation.

As Table 7.11 shows, the least adopted contractual arrangement is Lease. As in the previous cases, provision of water services at the national level is mostly a ministerial responsibility (generally the Ministry of Environment and Water Resources or the Department of Water Affairs) or that of the Central Government (Israel) rather than National Agencies (Japan and Malaysia). Municipalities have a lower level responsibility in China, India, Israel, Japan, Palestine and South Korea while Basin Commissions, Water Authorities, Provinces and Federal States act at an intermediate level of responsibility in Australia, China, India, Indonesia, Jordan, Lebanon, Malaysia and Thailand. Regulator supervises and monitor pricing policies only in Indonesia, Malaysia, Palestinian Territories and Saudi Arabia while in Australia alone full-cost-recovery issues are specifically included in the national legislation.

Table 7.11 Contractual arrangements of PPPs in Asia and Oceania.

Contractual Arrangements	Country	Incidence (%)
Management contracts	India, Japan, Lebanon, Malaysia, Saudi Arabia	63
Operation & Maintenance	Australia, India, Japan, Malaysia, Saudi Arabia	63
Lease (*Affermage*)	Thailand	13
Concession – BOT/DBO	China, Indonesia, Malaysia, Thailand	50
Joint venture	Australia, China, Indonesia, Saudi Arabia	50

7.3.4 Europe

Table 7.12 presents water and sanitation coverage as well as the percentage of these that are managed under a PPP agreement in Europe. The provision

Table 7.12 European countries water services: access and governance.

Country	Population (MM Inhab.)	Urban Popul. (%)	Access to Water (%)	Access to Sanitation (%)	Public Management (%)		PPP Management (%)	
					Water	Sanitation	Water	Sanitation
Austria	8.5	66	92	93	92	100	8	0
Belgium	11.2	98	100	86	97	90	3	10
Bulgaria	7.2	73	99	74	80	80	20	20
Croatia	4.3	58	93	76	100	82	0	18
Cyprus	0.8	67	100	66	100	65	0	35
Czech Republic	10.5	73	n.a.	n.a.	50	23	50	77
Denmark	5.6	87	96	90	99	100	1	0
England & Wales	56.6	82	100	97	0	10	100	90
Finland	5.5	84	91	82	90	100	10	0
France	66.0	79	100	82	33	53	67	47
Germany	80.8	75	99	97	79	80	21	20
Greece	11.0	77	n.a.	n.a.	65	62	35	38
Hungary	9.9	70	95	68	72	75	28	25
Iceland	0.3	94	n.a.	n.a.	90	n.a.	10	n.a.

Northern Ireland	1.8	63	98	76	99	60	1	40
Italy	60.8	69	98	87	58	68	42	32
Lithuania	2.9	67	74	62	100	100	0	0
Luxemburg	0.5	90	n.a.	n.a.	100	n.a.	0	n.a.
Malta	0.4	95	n.a.	n.a.	100	n.a.	0	n.a.
Netherlands	16.9	89	100	100	100	90	0	10
Norway	5.1	80	89	84	84	95	16	5
Poland	38.5	61	88	65	87	97	13	3
Portugal	10.5	62	96	82	75	76	25	24
Romania	19.9	54	57	44	89	89	11	11
Russia	146.1	74	n.a.	n.a.	90	94	10	6
Scotland & Ireland	7.5	89	97	83	100	n.a.	0	n.a.
Slovakia	5.4	54	87	60	80	80	20	20
Spain	46.5	79	100	98	51	49	49	51
Sweden	9.7	86	89	89	95	99	5	1
Switzerland	8,1	74	99	97	100	100	0	0
Turkey	76.7	72	n.a.	n.a.	98	90	2	10
	735.5	**74**	**93**	**81**	**82**	**78**	**18**	**22**

Note: population data refer to official and World Bank estimations for 2013–2014; other data refer to 2012.
Source: World Bank (2014a, b).

of water and sanitation services is totally privatized in England and Wales while France, Czech Republic, Spain and Italy present a high proportion of PPP management (*Source*: EUREAU, 2009).

National provision of water services is a ministerial responsibility (generally the Ministry of the Environment, in some case – Italy, Lithuania, Norway, Poland and Sweden – jointly with National Authorities) or that of in National Water Departments or Agencies (Croatia, Cyprus, England & Wales, Germany and Scotland). Municipalities have a lower level responsibility in almost every country (excluding Russia, England and Wales, Scotland) while Regional Governments or Directorates, River Basin and local Authorities, Provinces and Landers, act at an intermediate level of responsibility in Belgium, Bulgaria, Croatia, Czech Republic, Denmark, England and Wales, Finland, France, Germany, Italy, Netherlands and Spain.

Regulator supervises and monitor pricing policies in Bulgaria, England and Wales, Germany, Italy, Portugal, Romania, Scotland and Ireland, Slovakia while in Belgium, England and Wales, Finland, France, Germany, Italy, Netherlands, Norway, Portugal, Scotland and Ireland, Sweden and Switzerland full-cost-recovery issues are specifically included in the national legislation. Concession and joint ventures are the preferred PPP contractual arrangements (Table 7.13).

Table 7.13 Contractual arrangements of PPPs in Europe.

Contractual Arrangements	Country	Incidence (%)
Management contracts	Germany, Northern Ireland, Norway, Poland, Russia, Slovakia, Turkey	35
Operation & Maintenance	France, Germany, Hungary, Northern Ireland, Russia, Turkey	30
Lease (*Affermage*)	Czech Republic, France, Germany, Hungary, Poland, Russia, Slovakia, Spain, Turkey	45
Concession – BOT/DBO	Bulgaria, Czech Republic, Denmark, Finland, France, Germany, Greece, Hungary, Italy, Poland, Portugal, Romania, Russia, Slovakia, Spain, Sweden, Turkey	85
Joint venture	Czech Republic, Denmark, France, Germany, Greece, Hungary, Italy, Norway, Poland, Portugal, Russia, Slovakia, Spain	65

Table 7.14 Examined countries water services: access and governance.

Continent	No. of Countries	Popul. (M Inhab.)	Urban Popul. (%)	Access to Water (%)	Access to Sanitation (%)	Public Management (%)		PPP Management (%)	
						Water	Sanitation	Water	Sanitation
Africa	11	360.6	45	65	42	82	94	18	6
Americas	8	816.9	82	81	59	78	80	22	20
Asia and Oceania	15	3,144.3	49	89	68	84	92	16	8
Europe	31	735.5	74	93	81	82	78	18	22
	65	**5,057.3**	**58**	**82**	**63**	**82**	**86**	**18**	**14**

Note: population data refer to official and World Bank estimations for 2013–2014; other data refer to 2012.
Source: World Bank (2014a, b).

7.4 CONCLUSIONS

The tables above show that, as previously suggested, PPP experience in the water sector varies greatly from one country to another according to the different reactions Governments have had to the negative impact of the increased cost of debt during crisis periods. Table 7.14 presents water and sanitation coverage as well as the percentage of these that are managed under a PPP agreement in the examined countries worldwide highlighting that: on average 82% of 5 million inhabitants living in the 65 countries (58% of which in urban areas) have access to water while only 63% have access to sanitation; in these countries the public sector is the main provider supplying, in general, more than 80% of water services while the private sector runs almost 18% of water and 14% of sanitation under different PPP regimes.

Many countries adopt more than one type of PPP contractual arrangement, in several cases even in combination. Nevertheless nearly two thirds are

Table 7.15 Contractual arrangements of PPPs in the examined countries (%).

Continent	Management Contracts	Operation & Maintenance	Lease (*Affermage*)	Concession – BOT/DBO	Joint Venture
Africa	33	17	33	83	<1
Americas	25	38	25	63	38
Asia and Oceania	63	63	13	50	50
Europe	35	30	45	85	65
	38	**36**	**33**	**74**	**48**

concentrated in concession regimes. Joint ventures are significantly favoured in Europe and in Asia/Oceania while O&M contracts are the most preferred in Asia/Oceania (Table 7.15).

At the national level the provision of water services is mostly a ministerial responsibility (generally the Ministry of the Environment). Central African Governments or National American Agencies or European National Authorities may also have responsibility for water services. Municipalities have a lower level of responsibility in every country where Federal Governments, Provinces or local Authorities may also have an intermediate responsibility. Regulators operate in nearly one third of the studied countries while national legislations specifically include full-cost-recovery principles in a little less than 25% of the examined countries (mostly in the Americas and in Europe) (Parena, 2006).

REFERENCES

Brunet P. (2014). Peer performance Solutions: A Booming Innovative Contracting for Improved Utility Performance. IWA European Utility Management, Oslo 2014, IWA Publishing, London, UK.

EU Commission. (2009). Communication from the Commission to the European Parliament, the Council, the European Economic and Social Committee and the Committee of the Regions, Mobilising Private and Public Investment for Recovery and Long Term Structural Change: Developing Public Private Partnerships, EU Commission, Brussels.

EUREAU. (2009). Statistics Overview on Water and Wastewater in Europe 2008, Edition 2009, Country Profiles and European Statistics. EUREAU, Brussels, Belgium.

Garcia Quesada M. (2011). Water and Sanitation Services in Europe: Do Legal Frameworks provide for "Good Governance"? IHP-HELP Centre for Water Law, Policy and Science, Under the Auspices of UNESCO. University of Dundee, Scotland, UK.

IMTA/OECD. (2008). Expert meeting for a beneficial private sector participation in the water and sanitation sector, lessons learnt from Latin American country experience, 4–5 September, Mexico.

International Benchmarking Network for Water and Sanitation (IBNET). (2010). Database for water and sanitation utilities performance data, www.ib-net.org.

IWA. (2004). Specialist Group on Statistics and Economics. European Water Regulation Survey. Helsinki Meeting.

Mohajeri S., Knothe B., Lamothe D.-N. and Faby J.-A. (2003). EU Commission, Directorate-General for Research. Global Change and Ecosystems, Acqualibrium: European Water Management between Regulation and Cooperation, EU Commission, Brussels, Belgium.

OECD. (2000). Global Trend in Water Supply and Wastewater Financing and Management: Changing Roles for the Public and Private Sectors. OECD, Paris.

OECD. (2003). OECD Policy Briefs: Public-Private Partnerships in the Urban Water Sector, Public Affairs Division. OECD, Paris.

OECD. (2011a). Meeting the Challenge of Financing Water and Sanitation. Chapter 8: Tools and Approaches – Guidelines for Performance Based-Contracts. OECD Studies on Water, OECD, Paris.

OECD. (2011b). Meeting the Challenge of Financing Water and Sanitation. Chapter 10: Private Sector Participation in Water Infrastructures – Checklist for Public Action, OECD Studies on Water., OECD, Paris.

Parena R. (2005). Economics and financing in the water sector, Biological, Physiological and Health Sciences, Water and Health, In: Encyclopedia of Life Support Systems (EOLSS), Developed Under the Auspices of UNESCO, W. Grabow, (ed.), EOLSS Publishers, Paris.

Parena R. (2006). Water Regulation and International Charging Policies at a Glance. IWA World Water Congress, 10–14 September, Beijing, PR China, IWA Publishing, London.

Parena R. (2010). Worldwide Regulation and Tariffs Structure. IWA Utility Management Conference, Barcelona, IWA Publishing, London, UK.

UN Economic Commission for Latin America and the Caribbean. (2014). Network for Cooperation in Integrated Water Resources Management for Sustainable Development in Latin America and the Caribbean, Circular 40. UN Economic Commission for Latin America and the Caribbean, Santiago, Chile.

WHO/UNICEF. (2014). Joint Monitoring Programme, Progress on Drinking Water and Sanitation Report, Update 2014. WHO/UNICEF, Geneva, Switzerland.

World Bank. (2014a) Data on Urban Population, World Bank Development Indicators. World Bank, New York.

World Bank. (2014b). PPP Arrangements: Types of Public-Private Partnership Agreements, Public-Private Partnership in Infrastructure Resource Centre. World Bank, New York.

Chapter 8

Access to public water services

*Carlos Pereira and Alexandra Cunha**

The Water and Waste Services Regulation Authority (ERSAR), Rua Tomás da Fonseca, Torre G - 8º, 1600-209 Lisboa, Portugal

**Corresponding author Email: alexandra.cunha@ersar.pt*

ABSTRACT

This chapter aims to explain the role of regulation in the access to public water services as a tool of public policy. It consists of three sections: the first provides a major approach of the main characteristics of the sector of water services, which will allow the reader to get a perspective of its relevance, under several analysing prisms; the second seeks to contextualize the intervention of public authorities in the water services sector, in the sense that they are responsible for the definition of public policies that will boost the right conditions to ensure the provisions of these services to the population, and have the power to legislate and regulate how the sector is organized in each country, the principles to be taken into account in providing these services and the public service obligations inherent; and finally in the third we discuss the need for regulation in this sector, as a tool of public policy.

8.1 MAIN CHARACTERISTICS OF WATER SERVICES SECTOR

In this chapter, when we refer to water services we wish to refer to all activities related to the management of water resources for all possible uses, i.e.

 Water Consumption, Tariffs and Regulation
Editor(s): Francesc Hernández-Sancho
doi: 10.2166/9781780404684-8_0139

abstraction, storage, transport, treatment, distribution of surface or groundwater, as well as the collection, appropriate treatment of wastewater and subsequently discharge into surface water bodies. The water sector involves drinking water supply and wastewater management services which are crucial to the well-being of citizens, to public health, to economic activities and to the protection of the environment. This explains to a large extent why this sector should be guided by principles such as global access, continuity and quality of service, efficiency and equity of pricing.

In general, water services are known as public services that fulfil the basic needs of humans, provided in an exclusive way, through a network infrastructure which is mostly underground. It is a capital-intensive industry that requires significant 'up-front' financing for construction, although usually benefits from the support of non-reimbursable public funds in order to make viable investments and accessible services at socially affordable prices. However, these services have some very specific features that deserve special attention, such as: the way the sector is organized; the variety of governance models existing; and the permanent need to reach a balance between economic sustainability of management entities, economic accessibility for people and environmental sustainability. As a matter of fact, the water services sector can be characterized by being structurally complex, highly socially sensitive, economically relevant and environmentally very important, which makes it simultaneously engaging and challenging for those who work in it.

8.1.1 Complexity

The complexity in the water sector comes, first of all, from the fact that these services are essential and must be provided in continuity, respecting the legally defined quality standards and avoiding negative environmental impacts. Also, they can be assured directly by public authorities, holders of infrastructure or by public or private companies under delegation or concession contracts. Thus, some of these entities are governmental or municipal agencies, organized for purposes other than generating profit, that operate side by side with commercial companies, all of them delivering public services of water supply and sanitation. Moreover, since these services are based mainly in an underground infrastructure network, monitoring and maintenance require sophisticated technologies and highly trained technicians.

We should also mention the fact that public water services mostly cover the water supply and wastewater sanitation, each one involving a set of activities throughout its value chain, as shown below (Figure 8.1), and rarely the activity business of the utilities covers all value chain stages. Although there is a trend towards vertical integration of these services, which means that a single entity attempts to fulfil most or all value chain stages, activities above the yellow line in Figure 8.1 are often developed at a regional level, spanning several

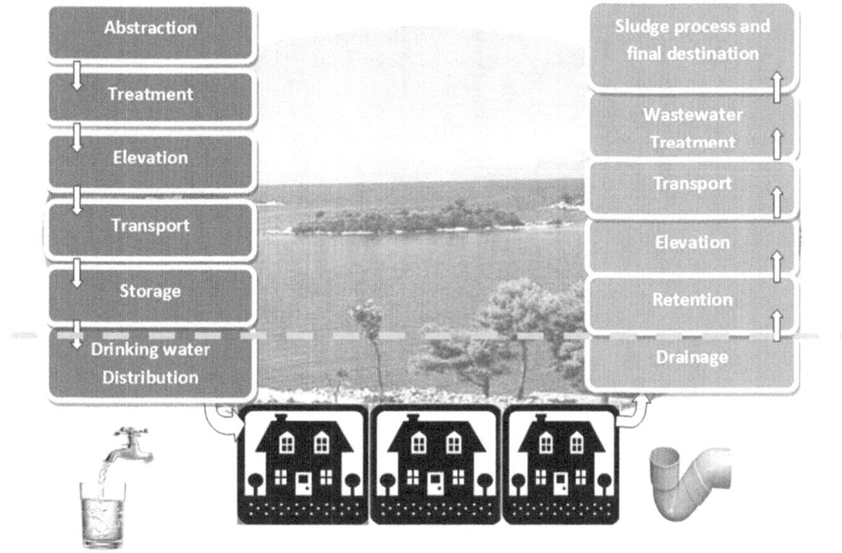

Figure 8.1 Chain value activities of water services.

municipalities (bulk systems), while below the line, activities are usually developed at a local level.

Making use of the data from an international comparison study by Cunha Marques (2010) regarding governance, we can observe it clearly confirms the trend towards the coexistence of public and private entities. The infrastructures are mostly owned by public authorities (see Table 8.1), while in terms of management systems we find public and private entities. Sometimes the municipality is the owner of the system and provides itself services to the end-users, otherwise services are provided by a public company on behalf of the public authority (delegation), or by private companies acting under a public service concession contract.

Despite the common coexistence of public and private entities in most countries, at a local level, services are still often provided by municipalities (direct management) or by local companies (delegation) to the end-users (retail services). In addition, some utilities operate only on one of the services – drinking water supply or wastewater sanitation – while others provide both services, seeking to gain economies of scope.

The information shown in the table below, regarding data collected for an international comparison, allows us to find the typical profile of governance in the water services sector, as a sector in which, in the vast majority of cases, infrastructure is owned by public authorities, sharing this responsibility between the central government and local authorities. The management and operation of

Table 8.1 Governance models in the supply sector and wastewater sanitation.

Countries	Assests Ownership (a)	Governance Models						
		Drinking Water System Management		Wastewater System Management		Vertical Integration (from Abstraction to Distribution to the End User)		
		Direct Management or Delegation in a Public Agency	Concession to Private Companies (b)	Direct Management or Delegation in a Public Agency	Concession to Private Companies	Y	N	Mixed (Both Situations Exist)
AUSTRALIA	Public (National and local authorities)	X		X				X
BELGIUM	Public (National and local authorities)	X		X	x			X
BRAZIL	95% public (National and local authorities) and 5% private	X	x	X	x			X
CANADA	Public (National and local authorities)	X	x	X	x	X		
CAPE VERDE	Public (National authority)	X	X	X	X	X		
CHILE	Mostly private		X		X	X		

COLOMBIA	Public (National and local authorities)	X	x	X	x	X	
COSTA RICA	Public (National and local authorities)	X	x	X	x	X	
CZECH REPUBLIC	Mostly prublic authority ownership	x	X	x	X	X	
FRANCE	Public (National and local authorities)	X	X	X	X		X
INDONESIA	Public (National and local authorities)	X	x	X	x		X
ITALY	Public (National and local authorities)	X	X	X	X	X	
MOZAMBIQUE	Public (National authority)	X	X	X	X	X	
PORTUGAL	Public (National and local authorities)	X	x	X	x		X
ROMANIA	Public (National and local authorities)	X	x	X	x	X	
SINGAPORE	Public (National authority)	X	x	X	x	X	
SWEDEN	Public (Local authorities)	X		X			X
THE NETHERLANDS	Public (National and local authorities)	X	x	X	x	X	

(*Continued*)

Table 8.1 Governance models in the supply sector and wastewater sanitation. (*Continued*)

Countries	**Assests Ownership (a)**	**Governance Models**						
		Drinking Water System Management		**Wastewater System Management**		**Vertical Integration (from Abstraction to Distribution to the End User)**		
		Direct Management or Delegation in a Public Agency	**Concession to Private Companies (b)**	**Direct Management or Delegation in a Public Agency**	**Concession to Private Companies**	**Y**	**N**	**Mixed (Both Situations Exist)**
UNITED KINGDOM	England and Wales: mostly private; Scotland and Northern Ireland: public authorities	X	x	X	x	X		
US AMERICA	Public (National authority)	X	x	X	x			X
ZAMBIA	Public (National authority)	X	x	X	x	X		

Source: Based on Cunha Marques (2010).

Notes: (a) In the case of concessions, when it includes the construction or/and maintenance of infrastructures, ownership is transferred to the State at the end of the concession period. (b) In legal terms, the prevailing contractual relationship is the concession agreement, but there are other situations, such as delegation in private and licensing also considered in this column.

water supply systems and wastewater treatment is still mainly ensured by the municipalities or by entities owned by the state and by local authorities, but it is common to also find intervention of private entities. Finally, the data in the table show also that in most countries these services are vertically integrated, that is, the same utility fills all stages of the value chain, from abstraction to the distribution to end users. It is also important to mention that the water sector is a typical case of network industry, and that we can find more than one utility with different corporate natures along the value chain. This displays the complexity of the sector, which naturally requires a specific approach by the public authorities.

8.1.2 Social sensitivity

Water supply and management is also a highly socially sensitive activity, since water and sanitation services are essential to ensure the social welfare of a people and is a key factor in preventing public health problems. Studies by the World Health Organization pointed to a strong link between water and health, and show the profound influence of water supply and quality on public health, concluding that in some regions of the world there are still a high percentage of deaths due to inadequate conditions of water supply and sanitation (see Figure 8.2).

Everyone all over the world requires clean and safe water as a basic need, as well as sanitation. Thus, water supply and sanitation services must be geographically

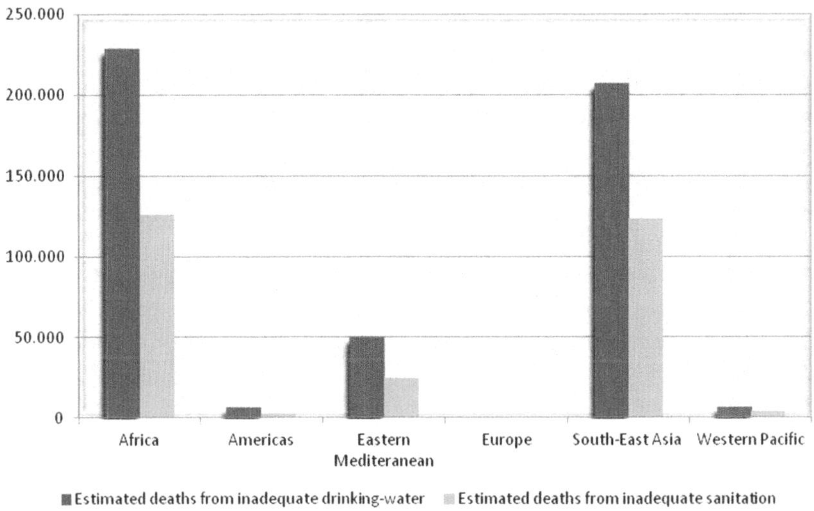

Figure 8.2 Estimated deaths from inadequate drinking water and sanitation in low- and middle-income countries by region, 2012. Source: World Health Organization (2014).

accessible and provided at affordable prices to the population, while utilities should be able to have economic and financial sustainability to ensure the continuity of the operation in an efficient productive base. On 28th July 2010, through Resolution 64/292, the United Nations General Assembly explicitly recognized the human right to water and sanitation and acknowledged that clean drinking water and sanitation are essential to the fulfilment of all human rights. The Resolution calls upon States and international organisations to provide financial resources, capacity building and technology transfer to help countries, in particular developing countries, to provide safe, clean, accessible and affordable drinking water and sanitation for all.

A true development process, in any country, has to address the need to make these services widely available across its territory, with acceptable standards concerning the quality of the service provided and the price. As noted by De Albuquerque and Roaf (2012), "*The rights to water and sanitation move universal access to water and sanitation from being a 'good idea' to being a legal entitlement. While ensuring the delivery of safe and affordable water and sanitation services is central to realising the rights to water and sanitation, it is equally essential that the various actors responsible for providing and monitoring those services are aware of their specific obligations under international and national law and are held accountable by users when they fail in their responsibilities*". Aspects such as physical access to services, water quality, the price and the quality of services are socially sensitive and have been of great importance for populations, and frequently are the subject of discussion in campaigns for local elections.

From another perspective, we must not ignore the accessibility of infrastructures for access tap to water and sanitation services. However, for cultural or economic reasons, people do not make the connection to the system, continuing to use unsecured precarious and temporary solutions. This type of situation should be prevented through proper planning of the investments based on studies to find out the real needs and to find out which are the best solutions in each case and, simultaneously, warning the population about the risks of consuming water without a guarantee of quality standards.

Assuming the infrastructures network for drinking water supply and sanitation were built because they correspond to the actual needs of the population, once they are available, all households served by this network must be connected, thus preventing overcapacity and ensuring the share of investment costs over the largest possible number of customers.

8.1.3 Economic relevance

Water supply is also an economically relevant sector. On one hand, regarding the consumed resources, the water sector is characterized by factors such as intensive capital and high investment payback periods, on the other hand, by its nature, water is not only vital to human life as well as necessary to ensure that agriculture, fisheries, transport, energy, tourism and many other industries develop and prosper. Also most industries need safe water for their production processes and adequate drainage systems at competitive prices.

According to European Commission paper (EC, 2011), "*the total turnover of industries active in the provision of environmental services (ecoindustries) in the EU-27 was EUR 319 billion in 2008, equivalent to 2.5% of the EU's GDP. The EU's environment industries represent around 3.4 million direct jobs. The most important sectors in terms of revenue are by far waste management, water supply, wastewater management and recycled materials*".

However, the weight of this sector in the economy is not limited to entities directly responsible for the management of water and sanitation systems. There is a wide range of other economic agents in this sector, which generates an intense economic activity. In addition to the entities directly involved in the management and operation of the systems – which, as stated above, can be ensured by municipalities, associations of municipalities, local companies, state-owned companies or private concessions – there are, at a Government level, services from the Public Administration, such as government departments responsible for monitoring and regulation of this sector of activity, environmental authorities and independent regulators, as well as other players with no intervention in the management of the systems, but playing an important role in the sector.

For instance, contractors, manufacturers and suppliers of materials, equipment and products, consultants and project companies, research and development centres, training centres, analysis and testing laboratories, financing entities and insurance companies. Furthermore, civil society is, of course, an essential player, through consumer protection associations, private-sector companies associations, associations for the protection of the environment and the public in general. Therefore, the impacts on this sector of the economy are much more expressive than the direct impact from the industry's activity in the provision of environmental services, because it mobilizes a large number of stakeholders with different legal natures and approaches, and a wide range of public organizations and private companies with different roles.

8.1.4 Environmental impacts

We must, finally, pay special attention to the environmental impacts of water services, either from the perspective of water as an essential and scarce good in most parts of our planet, or from the perspective of used water (wastewater) rejection, which must be handled in a sustainable way, preventing pollution of the environment. The United Nations World Water Development Report, 2015: Water for a sustainable world (WWAP, 2015), foresees a dramatic scenario for future decades, if the world population continues to grow and if drastic measures to reduce the consumption of drinking water are not taken.

It says: "*Unsustainable development pathways and governance failures have generated immense pressures on water resources, affecting its quality and availability, and in turn compromising its ability to generate social and economic benefits. The planet's capacity to sustain the growing demands for freshwater is being challenged, and there can be no sustainable development unless the balance between demand and supply is restored. By 2030, the world is projected to face a 40% global water deficit under the business-as-usual scenario*".

In summary, preserving water is preserving health and well-being, but it is also a way of ensuring that agriculture, fisheries, transport, energy, tourism and many other industries develop and prosper. However, especially in the more developed societies where water consumption per capita is higher, we need to reinforce measures in order to reduce demand and ensure a more sustainable use of our water.

Sales income from water services aims to finance this activity and ensure the future sustainability of the entities that provide such services, of course in an efficiency basis, but the way we design tariffs should also be seen as a tool to promote environmentally sound practices. So, in addition to all the education and awareness campaigns to alert population to the drinking water scarcity problem and induce more environmentally friendly behaviour, tariffs should be designed to allow affordability to consumption levels appropriate to the needs of human life, and penalize consumption considered excessive when compared to referential consumption patterns to preserve water resources and the environment. It is common nowadays to set tariff models with a fixed component, aimed at ensuring permanent service availability and a variable component proportional to consumption, in stepped blocks.

Some European States also apply charges and taxes on groundwater and surface water abstraction. They seem to have primarily a financing (rather than an incentive) purpose. In fact, there is no clear evidence on the effectiveness of the outcome of charges and taxes applied on groundwater and surface water abstraction over water-use efficiency, which emphasizes the importance of a pricing policy to induce good practices in water consumption.

8.2 THE NEED FOR PUBLIC POLICIES

No one can be deprived of drinking water provision and sanitation services, which must be provided at good quality and affordable prices, regardless of the technical solution that best fits each situation. Moreover, as we saw above, these services are also very relevant, from various points of view, to other economic activities. Water-supply and wastewater services are classified by public authorities in general, at national, regional or local level, as public services of general economic interest (SGEI), subjected to specific public service obligations. The European Commission has clarified that "*SGEIs are economic activities which deliver outcomes in the overall public good that would not be supplied (or would be supplied under different conditions in terms of objective quality, safety, affordability, equal treatment or universal access) by the market without public intervention*" (EC, 2011).

As a consequence, water services must fulfil a number of requirements such as the principle of universality, continuity, quality of service, equality of access, affordability and transparency. At the same time, the provision of water services needs important investments and also management and operating costs are involved. All these items need to be covered in the most efficient way to ensure its sustainability over time. Public authorities, at a national level, should define and promote appropriate public policies aimed at assuring access to these services, establishing sector development plans and ensuring their implementation. They have the duty to define strategic lines for these plans, set deadlines for the implementation of the actions planned and nominate the entities responsible, as well as create the conditions for the sector to be properly organized, in order to ensure economic and financial viability investments, so they can attract public or private investors.

Public authorities also play the key and irreplaceable role of publishing laws governing this activity, setting the general principles to be respected and all other aspects of contracting services and governance models, as well as the relationships with consumers. Furthermore, we must also take into account that the provision of water supply and wastewater services is often a responsibility of local public authorities, which have an obligation to ensure that such services are satisfactorily provided within their geographic area of competence. This responsibility can be delegated to a local public company or entrusted, through a public service concession, to a private company, which performs the services under legal and contractual conditions, for a limited period of time specified in the concession contract, usually in exclusivity.

Thus, public authorities play an important role as legislators and as decision-makers of public policies to the sector, but at the same time are directly involved in the market, as owners of the water supply and sanitation network systems, and often as utilities, either providing services directly by their own means, or through a public company. In these cases, a conflict of interests can emerge, to the extent that the same party establishes rules and simultaneously is a player who acts in the market like any other utility. Moreover, it likely contributes to more asymmetric information between the economic agents.

These arguments, namely in situations where imperfect competition happens such as monopoly, justify the need for a clear segregation between the public authorities charged with establishing public policies and the legal framework to the sector, from the economic agents directly involved in the sector as investors, service providers and consumers, and from the regulation of these services. That's why legal framework enforcement, implementation of public policies and regulation of the sector in general should be monitored or ensured by an independent entity, as it will be discussed in more detail in the following sections.

8.3 THE NEED FOR REGULATION

In the 1980s, we observed a trend change in the paradigm of the state role in the provision of public services. Until then, the state assumed the role of producer of essential public services. But, faced with high budget deficits and lack of capacity to meet the increasing investment required, it was forced to seek alternatives to attract new investors and look to enhance efficiency gains in production.

However, it was necessary to increase the regulation of these activities, now performed by other entities without the direct intervention of the state. In this context, theories of regulation emerged as a modern form of public intervention aimed at the protection of the public interest and seeking to guarantee a better balance between utility and consumer interests, in the provision of public services under more transparent processes, especially when market failures are identified. Public authorities should set policies for the sector, in particular through the approval of strategic plans for its development, defining an adequate legal framework and promoting research and development studies, as well as disclosing sound and reliable information, accessible to all interested parties.

Strategic plans play a very important role in the industry orientation, because they are the instrument that should define the development goals of the sector, the general planning investments and the estimating resources. It should also identify the responsible bodies for implementing the proposed measures and the appropriate timing, as well as those who are in charge of monitoring and evaluating the developed actions and progress accruing, which is typically assigned to the regulator. The combined efforts of all parties are essential for the plan to succeed and obviously, all these aspects will be taken into account in the regulatory approach of the sector.

The regulator should be able to collaborate with the government in the implementation of public policies for the sector, having as guiding principles for its action the protection of the interests of consumers while safeguarding the sustainability conditions of the utilities and the promotion of balance with the objectives outlined in the strategic plan and with environmental and tariff policies in place. Thus, we must see regulation as an instrument of public policy, technically specialized and equidistant from all parties involved, whether public or private, municipal or state holders.

8.4 CONCLUSIONS

Water services constitute a typical case of a network industry, since the management of these systems tends to create natural monopoly situations. The main characteristic of a natural monopoly is that the optimal scale economies, regarding all relevant production levels, are achieved when there is only one utility. Therefore, the water utilities develop their activities often exclusively in a limited geographic area. The absence of a competitive market raises the risk of dominance by the utilities on consumers. This may lead to inefficiency and the result will probably be increased prices and a decline in quality, which should be avoided at all costs because they are public services essential to life and social welfare.

These features justify the need for a strong regulatory framework, whose goal is to create conditions of healthy and fruitful competition, similar to those that mostly exist in a competitive market. Thus, the regulatory framework should establish clear rules on public service obligations and criteria for setting tariffs; also it should define quality of service indicators to be observed by utilities, whatever governance model is adopted, regardless of its public or private nature. Such rules have an underlying concern in the protection of users, in terms of physical accessibility to services at affordable prices and with acceptable quality, while safeguarding the continuity of the provision of services through an appropriate economic and financial sustainability of the utilities.

REFERENCES

Cunha Marques R. (2010). Regulation of Water and Wastewater Services: An International Comparison, Vol. **9**, IWA Publishing. DOI: https://doi.org/10.2166/9781780401492

De Albuquerque C. and Roaf V. (2012). On the Right Track: Good Practices in Realising the Rights to Water and Sanitation. World Water Council, Lisbon, Portugal, 222p.

EC. (2011). Commission staff working paper, "Application of EU State Aid rules on Services of General Economic Interest since 2005 and the Outcome of the Public Consultation", 23 March 2011, 397, European Commission, Brussels, Belgium.

WHO (World Health Organization). (2014). Preventing Diarrhoea through Better Water, Sanitation and Hygiene: Exposures and Impacts in Low- and Middle-Income Countries. Switzerland. ISBN 978 92 4 156482 3.

WWAP (United Nations World Water Assessment Programme). (2015). The United Nations World Water Development Report 2015: Water for a Sustainable World. UNESCO, Paris.

Chapter 9

Economic regulation of water sector

Carlos Pereira and Alexandra Cunha

The Water and Waste Services Regulation Authority (ERSAR), Rua Tomás da Fonseca, Torre G - 8º, 1600-209 Lisboa, Portugal

Corresponding author Email: alexandra.cunha@ersar.pt

ABSTRACT

This final chapter aims to explain the economic regulation of water services safeguarding the interests of consumers while ensuring economic and financial sustainability of utilities. The chapter consists of six sections covering: a brief reflection on the benefits of an independent regulator; the main objectives of regulation in the context of public services, usually classified as services of general economic interest (SGEI), and therefore subject to specific public service obligations and fundamental principles; a possible approach to the regulatory framework of water services; highlights of some of the more relevant practical aspects of economic regulation, based on the experience acquired over five years as a member of the board of directors of ERSAR, the Portuguese Water and Waste Services Regulation Authority; a practical view of the water sector regulation in Portugal; and some concluding remarks.

9.1 WHY AN INDEPENDENT REGULATOR

According to the OECD (2014), "*Market failures and critical features of the water sector justify the establishment of high quality regulatory framework to serve the*

Editor(s): Francesc Hernández-Sancho
doi: 10.2166/9781780404684_0153

public interest at least cost for all actors involved. There is a large spectrum of regulatory functions routinely performed in relation to water services. Regulation of water services is not only about tariff setting, it involves other functions, such as the setting and monitoring of standards for access to and quality of services, of efficiency incentives, collection of information and monitoring of performance, and the organisation of users' participation."

The regulator occupies a privileged facilitative position between the policy and operational levels of the industry, acting as a channel for balancing the legitimate concerns of utilities, whether public or private, with the rights of the wider user community. This requires a deep understanding of the technology used in the industry, financing models of the operation and investments, consumer protection laws, the quality standards, economic regulation mechanisms and the environmental constraints to preserve. Furthermore, changes in public policies or in economic or technological drivers often require new regulatory initiatives, and even the established regulatory authorities may need to review their approach and methods of working. In fact, the OECD (2007) recognized that "*the overall regulatory framework governing infrastructure sectors is a key lever which policy makers across this organization should use to provide efficient, reliable, safe and affordable public services*". Thus, the specialised knowledge and skills required by an effective regulator would be difficult to obtain within the context of public administration. In parallel, a clear division between regulator and government departments – and for the same reason equidistant from all other interests involved, public and private – increases the degree in which the actions of the regulator are likely to be perceived as neutral and transparent, and so boosts market confidence in the sector. Such confidence, however, must be supported by regulatory behaviour and standards that are credible and inspire trust in all the stakeholders.

9.2 MAIN OBJECTIVES OF THE REGULATION

As we saw before, drinking water supply and sanitation services are public services, usually classified as services of general economic interest (SGEI), and therefore subject to specific public service obligations and fundamental principles, such as: equality or universality – must be available to all under the same conditions, which generally takes the form of a requirement to provide a universal service at uniform prices within the area of operation of each utility and within acceptable standards of quality; continuity – service must be provided on an uninterrupted and regular basis (supply obligation); and transparency – users should have clear, sound and accessible information about the utility's operating rules, on how prices are set and about consumer rights.

However, these services are based on a supply or sanitation network system, which entails high fixed costs related to the amortization of investments, infrastructure maintenance and repairs when necessary. Regarding economic performance, due to the existence of high fixed costs, average costs are decreasing in the long run, which makes the coexistence of more than one supplier to the consumers in the same geographical area not economically viable. So, water services are public services in which one of the main characteristics is the need for high scale economics at all relevant production levels. Thus, in this context, the most efficient number of firms to operate is one, which typically corresponds to a natural monopoly.

From a market perspective, competition appears to be the optimal and efficient mechanism that encourages producers and resource suppliers to respond to price signals and consumer sovereignty. When we face a situation of lack of competition, we find lower incentives for utilities to increase efficiency and quality of service and prices are potentially higher than necessary. Thus, something must be done to avoid non-competitive behaviour, to protect consumers and prevent abuse of dominant position. So, the question that arises is how to ensure that these public services, which are provided under monopoly conditions, and thus without the competitive pressure on price or quality, are provided to the population in productive efficiency conditions, at fair prices and sustainable conditions, fulfilling rigorously specific public service obligations and respecting those fundamental principles.

There are several ways to prevent or mitigate abuse of dominant position and fight inefficiencies, including through a strong legislative framework, through a state agency in charge of regulating or through an independent regulatory body. We find good examples on each of these solutions, but when there are public and private entities operating side by side in the market, the public authorities do not seem to be in a position to have a totally impartial and equidistant role from all stakeholders. So, in our opinion, on one hand, it is essential to regulate the way services must be provided and how utilities should be organized to prevent the negative effects of monopoly power; on the other hand, a market with so many different economic agents as this, needs a technically skilled regulatory body to play this role impartially and equidistantly from all the interests involved to ensure compliance with the rules and to monitor the quality requirements at fair prices.

There are many different ways to approach regulation, depending on the sector of activity and on the specific characteristics of the sector. One strategy followed by regulators, for instance in Portugal, and in our opinion with good results, is to create a competitive environment that leads utilities to feel a pressure to be more efficient, improve or maintain high levels of quality and practice acceptable prices. This competitive environment can be created through mechanisms usually called regulation for the quality or sunshine regulation, combined with models of price regulation, which aim to set prices that cover only efficient costs, safeguarding the good quality of the services and the sustainability of the sector.

First, we are considering non-punitive instruments which allow the utilities performance to be visible to their users and to create a healthy competitive environment for quality. From these instruments, we must highlight two: evaluation of quality of service and quality awards. The evaluation of quality of service comprises the implementation of an appropriate and coherent quality of service system assessment, and its annual and systematic evaluation, and the subsequent dissemination of results to the users of these services and to the public in general. Since no one likes to look bad, those who have unsatisfactory results tend to strive to improve their performance to get a better grade the following year. The quality awards are based on the same quality of service system assessment, but this time giving a seal of quality to better performances, rather than penalizing the poorer ones. Past experiences in Portugal in water and solid waste services showed that the winning utilities like to display their achieved quality seals and therefore work hard in order to keep them. Second, it is essential to set clear rules within the economic regulation, in particular the model of tariff setting, whether in a more elementary form through the application of the full cost recovery principle to public entities that provide these services or by applying more sophisticated models such as the price-cap or revenue-cap.

9.3 THE ROLE OF THE WATER SUPPLY AND WASTEWATER SERVICES REGULATOR

Regarding water services sector and considering its relevance in multiple aspects, as mentioned before, regulatory intervention requires a holistic approach, taking in account industry sustainability together with environmental, social and economic sustainability (see Figure 9.1).

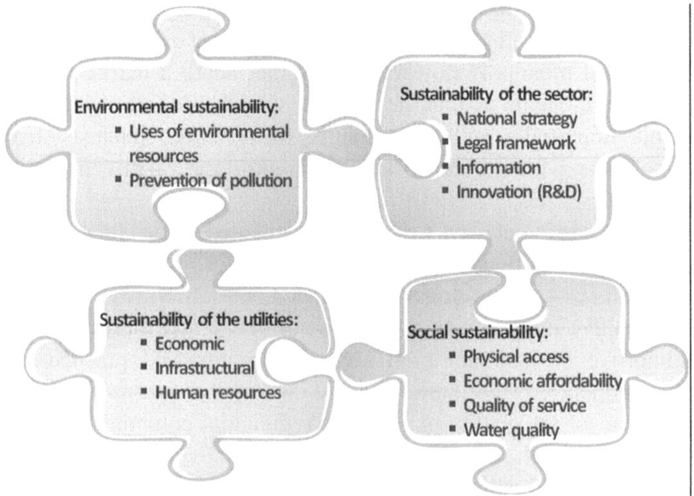

Figure 9.1 Sustainability of water supply and wastewater services under a holistic approach.

When we talk of environmental sustainability in the context of water services, we are referring mainly to the efficient use of drinking water and control of water losses, since, globally, water is a scarce resource. We are also referring to the prevention of the negative impacts that wastewater can have on the hydrological environment, especially if they are rejected without prior appropriate treatment and whether or not reused treated wastewater solutions are being considered. To reach a real effectiveness in this matter, the regulator should act in close cooperation with the environmental authorities of the country, which has an obligation to monitor and supervise the quality of the environment and has the power to impose sanctions when necessary. However, the regulator plays an important role too in environmental policies, such as establishing pricing models to facilitate access to water at affordable prices for low levels of consumption considered to be essential to life, and penalize abusive consumption, as well as losses of water higher than the acceptable levels. The regulator may also implement measures to encourage utilities to have environmentally friendly behaviour, penalizing situations not in accordance with the reference standards of the water losses benchmarks indicators.

The sector's responsiveness to the fulfilment of population needs in terms of safe drinking-water supply and sanitation depends largely on the existence of a public policy based on how to reach certain targets and also on what targets should be reached, as well as an appropriate strategy to implement them, and a consistent legal framework. The regulator should contribute to the formulation and implementation of these policies and to the definition of appropriate strategies, making suggestions and recommendations to the government in this area, as well as promoting research and technological development of the sector, and provide sound information to all interested parties. Furthermore, it is not only sufficient that water services are available in our homes. It is also necessary that the services should be provided to end users at affordable prices and within the quality standards of the services required, simultaneously safeguarding the sustainability of utilities.

In order to prevent the application of high costs, lack of productive efficiency and poor quality of services, which may lead to a decline in consumer welfare, the regulatory authority must pay special attention to the following aspects. (a) Consumer's rights and dispute resolution – in a monopolistic market, utilities often take advantage of their dominant position to impose on consumers some rules and prices that do not conform to legislation. In these situations the regulator must exercise their powers to not let it happen and provide easily accessible communication channels for consumers to present their claims. (b) Drinking water quality – subject to national and international quality parameters that utilities must comply, otherwise they can be sanctioned by the regulator. (c) Quality of service – if a service is provided in a monopolistic way, it may have little incentive to offer a good quality service. So, regulation should ensure that the public utility meets minimum standards of service, through an appropriate system of evaluation of the service quality. (d) Economic regulation – utilities have to be efficient in their production services, according to industry best practices identified by benchmarking within all industry costs in similar activities. Also prices need to cover the costs strictly necessary for the production of services and to remunerate the invested capital and not include inefficiency costs.

All these aspects fall within, what we call, behavioral regulation. But the regulator must also act on a structural level, in collaboration with the national and local authorities, particularly in the following areas: (i) public policy for the sector – cooperation with public authorities in defining public policies, pointing the way to reach certain targets to fulfill collective needs and defining the national strategy for the sector, namely proposing new legislation and improving the existing one, approving regulations and issuing recommendations; (ii) research, innovation and technical support – given the large number, diversity of scale and capabilities of utilities, the regulator should promote innovation and technical support, through cooperative partnerships with research centres and universities, promote edition of technical guides for the utilities and training events addressed to their employees; and (iii) reducing information asymmetry – collection, validation, processing and public disclosure of information, guaranteeing ease-of-use information for consumers.

Finally, we must keep in mind that a universal regulator must adopt a regulatory model to cover all entities, regardless of the governance model (direct management by public entities, delegations in state-owned or municipal-owned companies and concessions in private utilities) with the same level of requirements in terms of water quality, productive efficiency and service quality in general, and should be guided by an equidistance of interests. Nevertheless, given the variety of governance models we find in most countries, and considering the fundamental differences, according to their legal statute, their management rules and accounting system frameworks, the practical application of regulatory rules, particularly with regard to economic regulation, may need some adjustments.

9.4 SOME PRACTICAL ASPECTS OF ECONOMIC REGULATION OF THE SECTOR

In the context of economic regulation, the price of water and sanitation services assumes special importance because ultimately it provides the balance between economic accessibility for the population, economic sustainability of the utilities and the preservation of natural resources. The purpose of this section is to give an overall perspective of some relevant specific issues of economic regulation that greatly affect the price calculation. We have no pretention to make an exhaustive analysis of all economic regulation mechanisms and definitions of currently existing tariffs calculation models, with their advantages and disadvantages, because there is already a lot of good literature on this topic, written by renowned academics. As such, we have identified four key issues which will be discussed below.

9.4.1 Economies of scale

Although many exogenous factors determine the costs of providing water services, population dynamics can especially influence the level of water prices. As noted above, this is a network industry, so services are provided based on existing infrastructure that need economies of scale, at least sufficient to cover non-subsidized fixed costs, and to provide conditions for economic sustainability of supply and sanitation systems. Thus, when facing a reduction in demand, for demographic reasons or even for a decline in household incomes, the public utility does not have many management tools at its disposal to induce consumption increases through new customers. Also, it should not promote the increase of consumption per capita, because that would not be acceptable due to the general scarcity of drinking water.

In actual fact, negative population growth, which leads to a smaller customer base, will result in higher tariffs for those remaining customers, because the capital used for previous investment decisions must be amortised by fewer households and businesses. On the other hand, in general, a higher population density (or its increase) leads to cost reduction. For these reasons, strict planning and correct sizing of investments acquire a critical importance in this sector. Investments should be strictly dimensioned in accordance to the expected needs, according to relevant demographic studies, and operational cost structure must be as flexible as possible to allow the necessary adjustments over time. However, if the size of the demand does not provide enough interesting economies of scale so that the price is affordable to families, then we have to find economically viable alternative outlets, such as horizontal integration (aggregation of nearby villages) and/or vertical integration (concentration of the whole value chain on a single public utility).

9.4.2 Cost components

Users of water services should not pay costs other than those strictly necessary for the provision of these services. When a utility provides other services that have no relation to water services, as happens, for example, with municipalities that provide these services themselves through direct management, but also provide a wide range of other services to the community, a clear segregation of operational and investment costs is essential. The accounts of these entities must reflect a clear split between the water services and all other services that organization provides, and for regulatory purposes, the regulator should require that the accounts submitted are only those related to the regulated activity. As for the common costs to various activities, the entity shall internally adopt reasoned allocation criteria and use these criteria consistently over time.

Once bounded the costs involved in water supply and sanitation services, for the purpose of setting tariffs, relevant capital costs and necessary operating costs are accepted, including maintenance costs, depreciation and provisions if duly justified. Subsidies that the utility eventually receives and additional revenue eventually generated by the assets held for the provision of water services must be deducted and inefficient costs, identified according to previously defined criteria, should not be accepted. Two important components of the cost, depreciation and capital cost, are heavily dependent on the accounting standard practices and the assets ownership considered in the financial statements of the utility, and how the economic risk of asset maintenance and depreciation is allocated to the parties when system management and provision of services are entrusted to a third part under a concession.

For regulatory purposes, first it is essential to identify the relevant assets needed for the provision of services that will be paid for by the respective users. Second, the acquisition value of assets net of any non-refundable subsidies and other values which may represent a reduction of cost to the utility should be considered, since the tariff should only include the net capital for which the utility has the right of return. Third, it is important to clearly define the useful lifetime of the assets, which in the case of water and sanitation infrastructure is not a straightforward task, as they should always be in good operational conditions. However, there are engineering studies that help to determine the useful lifetime of these assets and there is also legislation on the rules for the calculation of depreciation for tax purposes. Finally, even for economic regulation purposes, it is also important to establish rules for improving investment and network expansion.

9.4.3 Drinking-water and sanitation-services pricing

Considering the industry's characteristics, already referred to above, in most countries, the prices of these services are not freely set by the public utility. In some countries, such as Germany, utilities can set their own prices, under the supervision of public authorities. But the most common situation is perhaps the one in which the prices, or maximum threshold of prices to be charged, are set by governmental authorities or independent regulators when they exist, relying on the principle of covering costs and the criteria of efficiency and sustainability. Nevertheless, there are cases where there are no technical criteria or the criteria are not clear.

In any case, the fundamental principle to be observed in drinking-water and sanitation-services pricing is the principle of cost recovery on an efficiency production basis. Following this principle, the average tariff to the end user is determined by dividing the total costs accepted, either by the total supplied water volume, or by the collected water volume if water use is metered, given that water tariffs for end users can take various forms, i.e., per household, depending on the areas of houses, or according to metered water consumption. However, the principle of cost recovery may imply a lack of incentives for the suppliers of water services to improve their efficiency. If they can be certain that their costs will be covered anyway, there is no need to be concerned about overinvestment or overcapacity. To prevent this from happening, we should ensure that the costs to be considered should only be the efficient ones, i.e. those that would be incurred by a service supplier behaving efficiently.

Given that this situation does not necessarily occur if demanding targets for efficiency and incentives to improve performance are not set, owing to a number of market imperfections along the value chain of water services (lack of competition, transaction costs, the weight of sunk costs and externalities), a regulator can play, in this matter, the important role of approving a Tariff Regulation, with clear rules and principles of setting tariffs and establishing an appropriate quality-assessment system based on efficiency and performance key indicators.

The regulator should set realistic targets in terms of efficiency, obtained through reasoned technical studies and benchmarking analysis, which will provide useful information for the decision on the costs to be considered for setting tariffs. Pricing rules should be clearly set out by the regulator and applied to all utilities, taking in account differences in legal nature and accounting standards, if relevant. Furthermore, they should be consistent over time, to ensure a desirable regulatory stability. The appropriate document to contain the rules for setting tariffs is the Tariff Regulation, which must be approved by the competent authority after consulting all stakeholders and analysing the suggestions and comments presented. Once approved, it should be widely publicized and explained, to facilitate understanding by utilities, consumers and other stakeholders.

Tariff Regulation should contain, at least, clear rules on the compensation model of capital invested net of subsidies (CAPEX), the qualifying criteria of the costs of acceptable operations for the purpose of tariffs (OPEX), and how efficiency goals are established and how they influence the price, which income resulting from the activity of the public utility is deductible in the costs, which rules apply to new investments and which incentives should be set to overcome the performance targets, for example, when the results lead to additional benefits for the community.

Of course the calculation formula becomes a bit more sophisticated when we consider capital costs, depreciation, incentives for efficiency gains and revenues from complementary activities. There are three main models for calculating tariffs often used in this industry, which are known as cost of service, price-cap and revenue-cap, and they all have their supporters and detractors. To some extent, their choice depends on the specific conditions of the sector operating in each country and also on the intrinsic relationship of these models with the way that people perceive the economic and financial risks associated to it. Not wishing to be exhaustive, we shall proceed to a brief description of each model, as we understand them.

9.4.3.1 Cost of service regulation (Rate-of-return)

This pricing method aims to reflect the total amount of money needed for a utility to operate and maintain its facilities, cover capital expenses, and provide an opportunity to earn a reasonable return. Cost of service regulation is often applied when the regulation is first implemented, the sector is growing, having not yet acquired maturity, and there is lack of data about the performance of utilities.

The main advantage of this economic regulatory mechanism is its simplicity and that it enables the regulated firm the opportunity to operate with a profit and attract capital for future growth. This revenue requirement then becomes the target revenue for setting prices. Nevertheless, the procedure is criticized for not providing strong incentives for cost containment, meaning that it does not include incentive tools for the regulated enterprise to become more efficient, namely reducing costs and improving service quality in an effective way to make a profitable investment.

9.4.3.2 Incentive regulation

While the cost of service regulation is based on a firm's actual costs, incentive regulation gives the regulator the option of setting prices based on the costs of an efficient firm or based on the costs regarding an international benchmark that reflects best practice. Within incentive regulation mechanisms, the best known are price-cap regulation and revenue-cap regulation, both aimed at providing strong incentives for the regulated firm to reduce its costs.

In a price-cap regulation framework, the regulator sets an initial price that should allow the regulated firm to cover its costs. Once the price-cap rule is established, the firm has more flexibility to determine how to operate its business in order to achieve maximum efficiencies and how to structure its prices (within the overall constraints of the price-cap framework) to reach its commercial objectives. The major advantages are the incentives for cost reduction and price stability over the price-cap period. The disadvantage of this regulation is that, after that period of time, relevant imbalances may arise resulting, for instance, from changes in demand and exogenous changes. If this happens, the public utility may be unable to attract new investment capital and the quality of their service may decline.

Revenue-cap regulation is another route for incentive regulation, and is based on the determination of the maximum-allowed revenues amount, and the average unit price, obtained by the quotient of this value vs. the expected volume. While price-cap regulation adjusts the public utility's prices according to thc price-cap index – that typically comprise the inflation impact on the operational inputs and the ability of the public utility to gain efficiencies relative to the average firm in the economy (called the X-factor) – revenue-cap regulation attempts to do the same thing, but for revenue rather than prices, setting the maximum allowable revenue that a utility can earn.

Revenue-cap regulation is more appropriate than price-cap regulation when the quantity demanded is not largely controlled by the regulated firm, consequently requiring annual adjustments in allowed revenue amount based on the cost factors that are influenced by that variation of the quantities. So the choice of revenue-cap rather than price-cap results in tariffs can be adjusted to compensate cost variations due to changes in demand, greatly reducing the risk of demand for the utility. This way, utilities will have less reason to complain about tariff changes to address any imbalances, greatly reducing the probability of the need for negotiation processes for economic and financial balance, which are commonly time-consuming, stressful and expensive.

Thus, the model becomes more complex, because in addition to the necessity of identifying a regulatory asset base and their variations, as well as the efficiency goals to be achieved, which are common to both mechanisms, the relevant cost drivers that justify adjustments in the tariff by the regulator have to be identified. However, this mechanism encourages the public utility to have a strong focus on efficiency and the benefits for the public utility and for end users may be very positive.

Since we often find various players along the value chain, each of them intervening only in part of the process, the rules for setting tariffs in the Tariff Regulation should be designed to be applied autonomously to each stage of the production process. We are thinking especially in situations where one entity manages a bulk system, providing services to other utilities in a regional area, and several utilities providing services directly to end users in a local level. We are not including mere service providers, who do not have direct responsibilities towards the public authorities.

Once the conceptual model for setting the average unitary tariff that the utility is allowed to charge is defined, it is important to ensure that consumers make efficient use of the water. There are many possible versions for water pricing and taxing systems to be applied to end users to achieve that objective without sacrificing the balance of utilities accounts. For consumers whose water use is metered, tariffs should comprise fixed and volumetric (per cubic metre) components, and the volumetric component should be designed to allow affordable consumption levels appropriate to the needs of human life, and penalize consumption considered to be excessive when compared with reference consumption patterns to preserve water resources and protect the environment. One of the most common systems is to raise block tariffs, which implies stepwise price increasing with increasing consumption, combined with a fixed fee.

9.4.4 Information to the consumer is a critical issue

Many people still think that nature has abundant fresh water and find it difficult to understand the price they have to pay for tap water. In fact, 99.7% of the Earth's water supply is not usable by humans. This unusable water includes saltwater, ice and water vapour in the atmosphere. Less than 1% of the worlds' freshwater, which is contained in rivers, lakes, and underground sources, can be used for human consumption. Before it can be delivered to the population, it needs analysis and treatment, and it also needs a network capable of transporting water in to homes. Furthermore, infrastructures need to be kept in a good condition.

The required investment and operating costs must be recovered, to ensure the future sustainability of utilities and ensure that the population will continue to have access to water services, unless the utility benefits from non-refundable subsidies granted by public authorities. However, if subsidies do not cover the full costs, as in most cases, then the remaining part of the costs have to be covered by tariffs. Of course, everyone understands this, but still, people often seem to be more willing to pay five hundred times more for bottled water than for tap water, despite this price being lower if compared with other public utilities.

In order to allow a better understanding by consumers, public authorities and utilities should provide clear and sound information about the service costs addressed to end users and to the interested public in general. Furthermore, consumers have the right to clear and accessible information on the quality of the tap water that is supplied to them, as well as on prices and on the quality of services that are provided.

Special attention must also be paid to the rights and obligations to consumers, keeping them informed and fulfilling the existing legislation. This is a very important issue commonly assigned to the regulatory body. In fact, the regulatory bodies have a duty to collect and disclose relevant and reliable information, and because they have a privileged access to information on all regulated utilities play a vital role in this regard.

9.5 CASE STUDY: TARIFF REGULATION IN PORTUGAL

We are currently in the process of altering the Tariff Regulation for the water sector in Portugal. As the sector is now much more mature and we have data about the performance of the utilities, it seems reasonable to move towards incentive regulation (as discussed before), namely price-cap and revenue-cap regulation. We opted to divide this section in two subsections. First, we will describe what prices the entities charge and whether they are profitable or not. Second, we will outline the economic relationship between end consumers and these tariffs. Our main goal is to give the reader a feeling for the Portuguese water sector, from the perspective of both the entities and end users through different sets of data.

9.5.1 Entities

In mainland Portugal, because some entities are vertically integrated and others are not, there is data for two tariffs: one paid by the service providers to the entities which own the bulk systems and another paid by the end consumers to the service providers. Until 2008, the values only reflect tariffs imposed by state-owned systems. Regarding the first set of tariffs, those paid the service providers to the entities which own the bulk systems, the average value in 2017 was 0.4920 and 0.5094 €/m^3 for the drinking water supply and wastewater management services, respectively.

Furthermore, there has been a slight increase in nominal terms over the years of the price set by these services. This increment reflects, in some cases, an increase in the awareness of the entities in terms of consolidation of investments. Nevertheless, in order to protect consumers, most systems still practise tariffs that are below what is necessary to cover all the costs. However, this increase is partially offset by a decrease in the tariffs in recent years. The reduction in 2015 was due to the effects of an aggregation of a large number of entities (Figure 9.2).

In 2017, the average prices paid by consumers to the service providers, which is obtained through the relationship between the total business volume (€/year) and the volume of total water supplied (m^3/year) for drinking water supply, and through the relationship between the total business volume (€/year) and the volume of wastewater treated (m^3/year) for the wastewater management services, were 1.28 €/m^3 and 0.87 €/m^3, respectively (Figure 9.3). Both drinking water supply and wastewater management services highlight a rather accentuated difference between their maximum and minimum price, with the companies charging higher prices. This discrepancy is mainly due to the fact that, for direct managed entities, not all the provision costs related to the service are recognized for accounting purposes in conjunction with the fact that the tariffs are usually below the average costs.

Given these tariffs, some entities are able to cover all costs and generate a positive aggregate net profit, while others are not able to perform as well. The degree of cost coverage is given in Figure 9.4 where UR is waste management services.

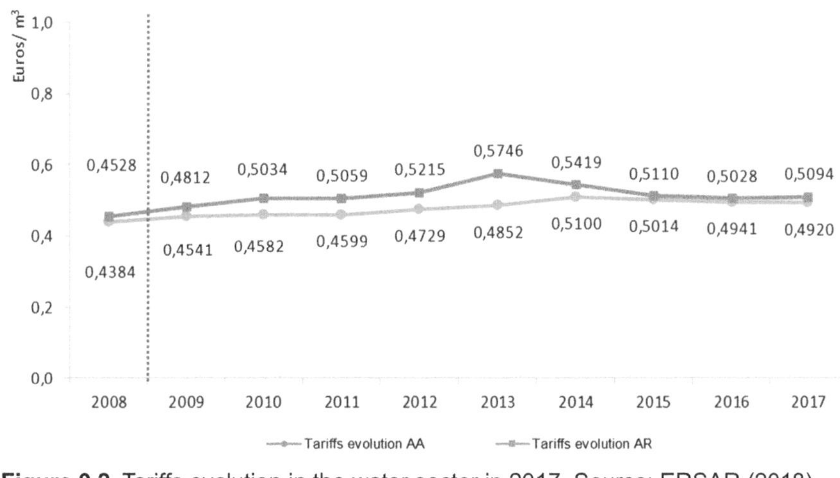

Figure 9.2 Tariffs evolution in the water sector in 2017. Source: ERSAR (2018).

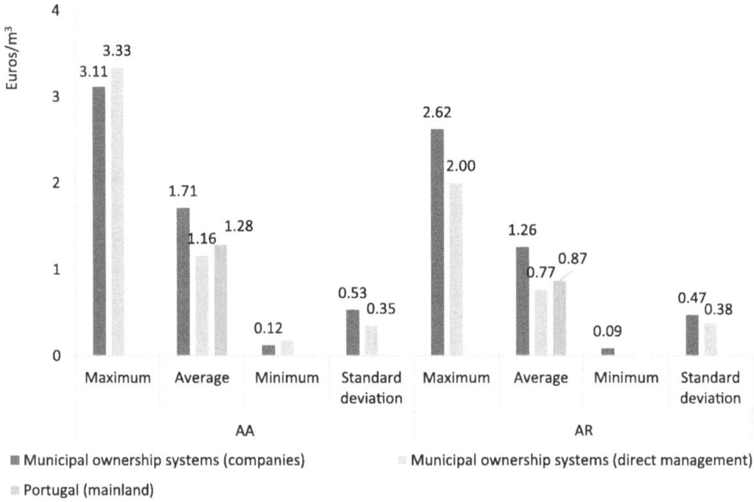

Figure 9.3 Average price in the water sector in 2017 for drinking water supply (AA) and wastewater management services (AR). Source: ERSAR (2018).

It is clear that entities who supply drinking water are able to cover more of their costs when compared with other entities. However, not all perform the same. Companies perform better, followed by the state-owned entities. The worst performers are the direct-managed entities, which cannot cover their costs in neither of the branches of the water sector.

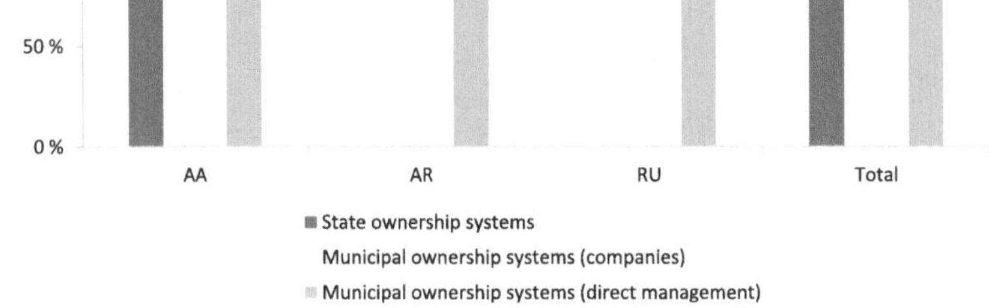

Figure 9.4 Degree of cost coverage in the water and waste management sectors in 2017 for drinking water supply (AA), wastewater management services (AR) and urban waste management services (UR). Source: ERSAR (2018)

9.5.2 Consumers

The aim of this section is to analyse the impact of these tariffs on consumer's lives in Portugal. Unfortunately, some of the data aggregates urban waste management. Nevertheless, a valid analysis can be conducted without loss of generality. First and foremost, it is clear by looking at Figure 9.5 that there is a positive relationship between the monthly income in a household and the monthly tariff paid by the same household. However, this correlation is not that strong (0.46), which indicates that a household's monthly income does not play a predominant role in the definition of the tariffs by the entities. As a matter of fact, the key factor when deciding on the tariffs is the total cost of providing the different services.

Moreover, there is no causality between the two variables given that $R^2 = 0.1274$. Therefore, a region with a higher monthly income per household does not necessarily have a larger amount of tariffs to pay per month, as depicted in Figure 9.6.

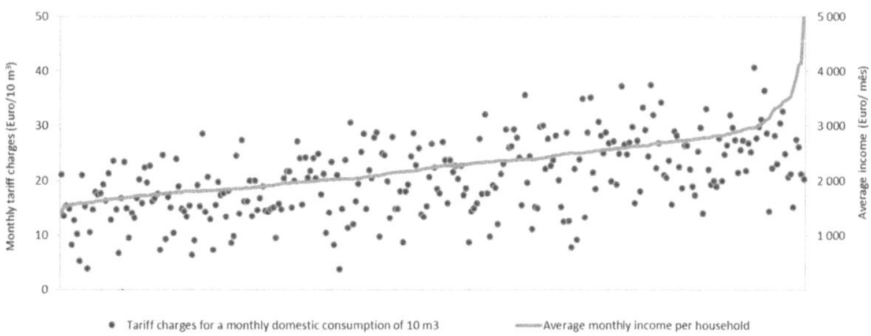

Figure 9.5 Monthly domestic tariffs and average monthly income per household in Portugal in 2017. Source: ERSAR (2018).

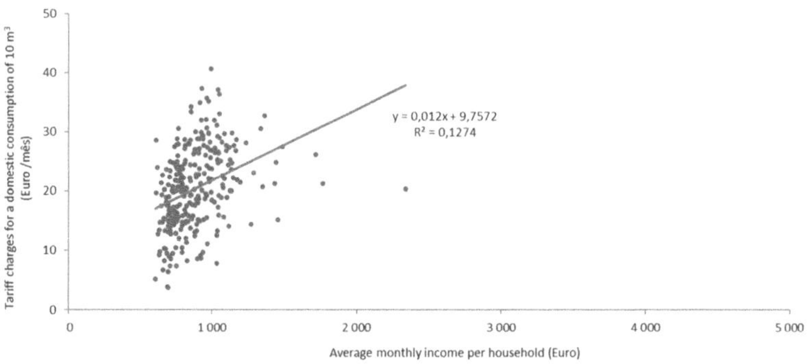

Figure 9.6 Scatterplot and regression line: Average monthly income per household versus tariffs for a domestic consumption of 10 m^3. Source: ERSAR (2018).

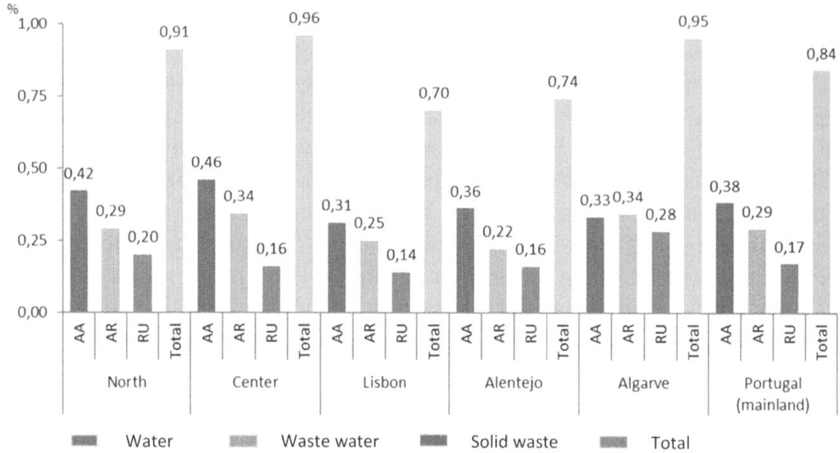

Figure 9.7 Average level of affordability in each service and region in Portugal in 2017. Source: ERSAR (2018).

One of the objectives as a regulator is for the consumers to get access to services at a reasonable price. Therefore, the affordability indicator (Figure 9.7) helps us accessing if that is indeed the case. In the water sector, this indicator is given by the monetary weight of these services on the household's income.

Overall, each one of the services presents good affordability with the wastewater management sector performing slightly better overall in this regard. This variable is not constant throughout the regions. For instance, the North, Center and Algarve regions are the worst in affordability for the three services, while in Lisbon and Alentejo are the best. This is expected as the different regions differ in various aspects such as total amount of population, population density, average monthly income, distribution of groundwater, which will in turn affect this indicator.

9.6 CONCLUDING REMARKS

Tap water supply and wastewater management services are crucial to the well-being of citizens, to public health, economic activities and to the protection of the environment, and hence should be guided by principles such as global access, continuity and quality of service and efficiency and equity of pricing. Furthermore, on 28th July 2010, through Resolution 64/292, the United Nations General Assembly explicitly recognized the human right to water and sanitation, and acknowledged that clean drinking water and sanitation were essential to the realisation of all human rights, which required governments to adopt measures for its implementation.

The provision of water services needs important investments, which often requires the attraction of private investors to develop infrastructures in partnership with public authorities. However, that would not be possible if there was no expectation of return on the capital invested, with adequate remuneration. Nevertheless, from a market structure perspective, the water sector is a typical case of a network industry, since the management of infrastructures tends to create natural monopolies, which naturally requires a specific approach by public authorities in order to prevent abuses by the dominant position. Therefore, public policies for the sector are needed, particularly in the approval of strategic plans for development and in a proper legal framework.

A regulatory body, in its turn, can play a very important role in monitoring and boosting those public policies, as well as promoting water quality and services quality assessments, setting rules for economic regulation proposes and tariffs, receiving and analysing complaints and providing the most appropriate follow-up, promoting research and development studies to help training in the sector and finally, collecting and disclosing sound and reliable information accessible to all the interested parties. For the reasons that we presented in this chapter, we believe a skilled and institutionally independent regulatory body, either from government or private interests, can be a strong instrument for public policy. The more the regulatory decisions and practices reveal a balance and an equidistance from all interests, the more effective is its action and the better the results will be.

REFERENCES

ERSAR. (2018). Relatório Anual dos Serviços de Águas e Resíduos em Portugal (RASARP), Volume 1 – Caraterização do setor de águas e resíduos. ERSAR, Lisbon, Portugal. (In Portuguese).

OECD. (2007). The Regulation of Public Services in OECD Countries: Issues for Discussion, Nov 2007. OECD, Paris.

OECD. (2014). Paper from the Regulatory Policy Committee, "Applying better regulation in the water service sector, The governance of water regulators". Meeting of the Network of Economic Regulators, OECD 5 November 2014, OECD, Paris.

Index

IWA Publishing's authorised EU representative for General Product Safety Regulations is Diane D'Arras, 15 rue Duret, 75116 Paris, France, e-mail: safety@iwap.co.uk.

Printed and bound by CPI Group (UK) Ltd, Croydon, CR0 4YY
12/05/2026
02108867-0014